STANDARDS, CONFORMITY ASSESSMENT, AND TRADE

Into the 21st Century

International Standards, Conformity Assessment,
and U.S. Trade Policy Project Committee

Board on Science, Technology, and Economic Policy

National Research Council

NATIONAL ACADEMY PRESS
Washington, D.C. 1995

NATIONAL ACADEMY PRESS • 2101 Constitution Avenue, N.W. • Washington, D.C. 20418

NOTICE: The project resulting in this report was approved by the Governing Board of the National Research Council (NRC), whose members are drawn from the councils of the National Academy of Sciences (NAS), the National Academy of Engineering (NAE), and the Institute of Medicine (IOM). The members of the expert committee responsible for the report were chosen for their special competencies and with regard for appropriate balance.

The report has been reviewed by individuals other than the authors according to procedures approved by a Report Review Committee. This committee consists of members of the National Academy of Sciences, the National Academy of Engineering, and the Institute of Medicine.

The Policy Division of the NRC consists of the Board on Science, Technology, and Economic Policy (STEP), the Committee on Science, Engineering, and Public Policy, and the Government-University-Industry Research Roundtable. The STEP Board reports as a unit of the Policy Division to the NRC Governing Board. This is the body by which the NAS, NAE, and IOM govern the work of the National Research Council.

This study was supported by the U.S. Department of Commerce's National Institute of Standards and Technology.

Library of Congress Cataloging-in-Publication Data

National Research Council (U.S.). International Standards, Conformity
    Assessment, and U.S. Trade Policy Project Committee.
      Standards, conformity assessment, and trade: into the 21st century
  / International Standards, Conformity Assessment, and U.S. Trade
  Policy Project Committee, National Research Council.
      p. cm.
    Includes bibliographical references and index.
    ISBN 0-309-05236-X
      1. Quality control—Standards—United States. 2. Manufactures—
  Quality control—Standards—United States. I. Title
  TS156.N366   1995
  658.5′62′021873—dc20   95-1649
                                                                  CIP

Printed in the United States of America

MEMBERS OF THE INTERNATIONAL STANDARDS, CONFORMITY ASSESSMENT, AND U.S. TRADE POLICY PROJECT COMMITTEE

GARY C. HUFBAUER, *Chairman*, Reginald Jones Senior Fellow, Institute for International Economics, Washington, D.C.

DENNIS CHAMOT, Associate Executive Director, Commission on Engineering and Technical Systems, National Research Council, Washington, D.C.

LEONARD FRIER, President, MET Laboratories, Inc., Baltimore, Maryland

STEVEN R. HIX, Chairman and CEO, Sarif, Inc., Vancouver, Washington

IVOR N. KNIGHT, President, Knight Communications Consultants, Clarksburg, Maryland

DAVID C. MOWERY, Associate Professor, Walter A. Haas School of Business, University of California, Berkeley

MICHAEL M. O'MARA, Business Leader, GE Plastics Cycolac Business, General Electric Company, Washington, West Virginia

GERALD H. RITTERBUSCH, Manager, Product Safety and Environmental Control, Caterpillar, Inc., Peoria, Illinois

RICHARD J. SCHULTE, Senior Vice President, Laboratories, American Gas Association, Cleveland, Ohio

SUSAN C. SCHWAB, Director, Corporate Business Development, Motorola, Inc., Schaumburg, Illinois

MICHAEL B. SMITH, President, SJS Advanced Strategies, Washington, D.C.

LAWRENCE L. WILLS, IBM Director of Standards, IBM Corporation, Thornwood, New York

Professional Staff

JOHN S. WILSON, Project Director
JOHN M. GODFREY, Research Associate
PATRICK P. SEVCIK, Project Assistant

v

The National Academy of Sciences (NAS) is a private, self-perpetuating society of distinguished scholars in scientific and engineering research, dedicated to the furtherance of science and technology and their use for the general welfare. Under the authority of its congressional charter of 1863, the Academy has a working mandate that calls upon it to advise the federal government on scientific and technical matters. The Academy carries out this mandate primarily through the National Research Council, which it jointly administers with the National Academy of Engineering and the Institute of Medicine. Dr. Bruce M. Alberts is President of the NAS.

The National Academy of Engineering (NAE) was established in 1964, under the charter of the NAS, as a parallel organization of distinguished engineers, autonomous in its responsibilities for advising the federal government. Dr. Robert M. White is President of the NAE.

The Institute of Medicine (IOM) was chartered in 1970 by the National Academy of Sciences to enlist distinguished members of appropriate professions in the examination of policy matters pertaining to the health of the public. In this, the Institute acts under both the Academy's 1863 congressional charter responsibility to be an adviser to the federal government and its own initiative in identifying issues of medical care, research, and education. Dr. Kenneth I. Shine is President of the IOM.

The National Research Council was organized by the National Academy of Sciences in 1916 to associate the broad community of science and technology with the Academy's purposes of furthering knowledge and advising the federal government. Functioning in accordance with general policies determined by the Academy, the Council has become the principal operating agency of both the National Academy of Sciences and the National Academy of Engineering in providing services to the government, the public, and the scientific and engineering communities. The Council is administered jointly by both Academies and the Institute of Medicine. Dr. Bruce M. Alberts and Dr. Robert M. White are chairman and vice chairman, respectively, of the National Research Council.

Preface

Product and process standards, as well as methods to ensure conformance to
these standards, have important implications for economic progress and public welfare. They also are increasingly important to global commerce. We hope
this book will serve as a reference document for public policy. It begins with a
discussion of the relationship between standards, product testing, certification,
and world trade. The volume then examines the role and responsibilities of U.S.
government and industry in the system. Emerging trends in key international
policies and programs are also addressed. The report concludes with a set of
recommendations both to strengthen the U.S. domestic system and to enhance
U.S. interests in overseas markets.

The National Research Council of the National Academies of Science and
Engineering was asked by Congress in P.L. 102-245 to study these issues (Appendix B). The Council's Science, Technology, and Economic Policy Board
provided the forum through which the study was initiated. A panel of experts
provided oversight of the resulting study and the professional staff work which
produced the final report.

The report addresses an extremely important set of goals for national policy.
These involve removing ineffective and duplicative rules and regulations that
govern testing, certification, and laboratory accreditation. Urgent reform is
needed in national conformity assessment policy. This will come about, in part,
through changes in the mandate of the National Institute of Standards and Technology. This report also discusses ways in which the United States can promote
open trade by removing standards-related barriers to trade and mechanisms to
better support U.S. exports in world markets. The U.S. should aggressively

eliminate barriers to global trade embedded in discriminatory foreign policies and practices. At the same time, we should lead the international community in creating a global network of mutual recognition agreements by governments with differing national conformity assessment systems.

Numerous individuals provided advice and assistance throughout the project. Most importantly, John Godfrey and Patrick Sevcik deserve great credit for their outstanding work. The committee served with extraordinary dedication to the success of this effort. Many individuals in government provided assistance to the project, especially those at the National Institute of Standards and Technology and the Office of the U.S. Trade Representative. Numerous experts in industry and universities also provided briefings, important information, and other assistance in our work. This is particularly true of those affiliated with the American National Standards Institute and other U.S. standards bodies.

Gary Clyde Hufbauer
Chairman

John Sullivan Wilson
Project Director

Contents

STANDARDS,
CONFORMITY ASSESSMENT,
AND TRADE

Executive Summary

The United States is the most productive and competitive nation in the world.[1] This fact is based on a high degree of efficiency in the domestic economy. In particular, significant progress has been made over the past several decades to foster a competitive economic environment for workers and firms. Initiatives by both industry and government to restructure the nation's productive capacities and promote microeconomic efficiencies have resulted in many benefits. This includes an acceleration of technological advance. We have eliminated many unnecessary rules and regulations that block U.S. firms and workers from taking full advantage of our creativity, industrial infrastructures, and technological edge. The United States has led the world in removing regulatory controls in the transportation, energy, and telecommunication sectors, for example. Continued progress, however, is needed if we are to move forward into the twenty-first century and achieve higher levels of productivity and economic growth. This progress will come, in part, through aggressive and targeted efforts to remove the remaining costly, inefficient, and unnecessary barriers to industrial production embedded in the U.S. national standards and conformity assessment system.

As we approach the year 2000, national welfare and economic strength will also increasingly center on the advantages the United States enjoys in global commerce. In addition to reform of the domestic economy, we need ever more innovative methods to promote goods and services overseas. The U.S. government must also continue to exercise leadership in the international community by aggressively removing the remaining barriers to trade. *A high-level focus by government and industry on standards and conformity assessment policy is one way of reaching these goals and promoting a more productive national economy.*

This report offers a comprehensive analysis of these subjects and the relationships among industrial production, standards, and conformity assessment. It provides recommendations to support both domestic policy reform, and the continued success of U.S. products in global markets. The information and data presented here support the conclusion that *in most instances, the U.S. standards development system serves the national interest well.* There is, however, evidence to indicate that our *domestic policies and procedures for assessing conformity of products and processes to standards require urgent improvement.*

At the same time, we must recognize the strategic importance of standards and conformity assessment systems in supporting national trade objectives. In order to address the new international dynamics of global trade, *an innovative U.S. trade policy to meet challenges of the post-Uruguay Round trading environment is required.* This should involve an integrated strategy by the U.S. government to link standards, conformity assessment, and trade. Our policies should aggressively seek to reduce standards-related barriers to trade. This involves both unilateral action through U.S. trade law and a new commitment to international negotiation aimed at mutual recognition by governments of conformity assessment systems.

The following summarizes the report's conclusions and recommendations, which are outlined in detail in each chapter of the report. An extensive discussion of the implications of these recommendations is included in Chapter 5.

CONFORMITY ASSESSMENT

The U.S. conformity assessment system has become increasingly complex, costly, and burdensome to national welfare. Unnecessary duplication and complexity at the federal, state, and local levels result in high costs for U.S. manufacturers, procurement agencies, testing laboratories, product certifiers, and consumers.

Government agencies should retain oversight responsibility for critical regulatory and procurement standards in areas of public health, safety, environment, and national security. The assessment of product conformity to those standards, however, is performed most efficiently and effectively by the private sector. Government should act only in an oversight capacity. The government should evaluate and recognize private-sector organizations that are competent to accredit testing laboratories, product certifiers, and quality system registrars.

- **RECOMMENDATION 1:** Congress should provide the National Institute of Standards and Technology (NIST) with a statutory mandate to implement a government-wide policy of phasing out federally operated conformity assessment activities.

  NIST should develop and implement a National Conformity Assessment

System Recognition (NCASR) program. This program should recognize accreditors of (a) testing laboratories, (b) product certifiers, and (c) quality system registrars. By the year 2000, the government should rely on private-sector conformity assessment services recognized as competent by NIST.

- **RECOMMENDATION 2:** NIST should develop, within one year, a ten-year strategic plan to eliminate duplication in state and local criteria for accrediting testing laboratories and product certifiers. NIST should lead efforts to build a network of mutual recognition agreements among federal, state, and local authorities.

After 10 years, the Secretary of Commerce should work with federal regulatory agencies to eliminate remaining duplication through preemption of state and local conformity assessment regulation.

STANDARDS DEVELOPMENT

The U.S. standards development system serves the national interest well. In most cases, it supports efficient and timely development of product and process standards that meet economic and public interests. Federal government use of the standards developed by private standards organizations in regulation and public procurement has many benefits. These include lowering the costs to taxpayers and eliminating the burdens on private firms from meeting duplicative standards in both government and private markets. Although not every public standard can be developed through private-sector processes, government should rely on private activities in all but the most vital cases involving protection of public health, safety, environment, and national security.

Current efforts by the U.S. government to leverage the strengths of the private U.S. standards development system, as outlined in the Office of Management and Budget (OMB) Circular A-119, "Federal Participation in the Development and Use of Voluntary Standards," are inadequate. Effective, long-term public–private cooperation in developing and using standards requires a clear division of responsibilities and effective information transfer between government and industry. Improved institutional mechanisms are needed to effect lasting change.

- **RECOMMENDATION 3:** Congress should enact legislation replacing OMB Circular A-119 with a statutory mandate for NIST as the lead U.S. agency for ensuring federal use of standards developed by private, consensus organizations to meet regulatory and procurement needs.

- **RECOMMENDATION 4:** The director of NIST should initiate formal negotiations toward a memorandum of understanding (MOU) between NIST and the American National Standards Institute (ANSI). The MOU should outline modes of cooperation and division of responsibility between (1) ANSI, as the organizer and accreditor of the U.S. voluntary consensus standards system and the U.S. representative to international, non-treaty standard-setting organizations and (2) NIST, as the coordinator of federal use of consensus standards and recognizing authority for federal use of private conformity assessment services. NIST should not be precluded from negotiating MOUs with other national standards organizations.

In addition, all federal regulatory and procurement agencies should become dues-paying members of ANSI. Dues will support government's fair share of ANSI's infrastructure expenses.

INTERNATIONAL TRADE

Expansion of global trade is increasingly important to domestic economic growth, productivity, and high-wage employment opportunities in the United States. The reduction of barriers to international commerce and aggressive promotion of U.S. exports must continue to be the fundamental objectives of a post-Uruguay Round trade strategy. At the multilateral level, the Uruguay Round of the General Agreement on Tariffs and Trade (GATT) achieved significant progress in reducing barriers related to discriminatory standards and national product testing and certification systems.

There is evidence to indicate that the growing complexity of conformity assessment systems in many nations threatens, however, to undermine future global trade expansion. U.S. exporters face high costs in gaining product acceptance in multiple export markets. Many nations impose duplicative, discriminatory requirements for product testing, certification, and quality system registration. The European Union's (EU's) mechanisms for approving regulated products, in particular, continue to pose serious barriers to expanded export opportunities for U.S. firms. Clearly, the severity of these obstacles varies by industry sector. From a national perspective, it is important, however, to achieve a rapid, negotiated removal of EU barriers. This will serve both to expand trade opportunities with our European partners, and to help promote the success of similar negotiations between the United States and other trading partners, especially those in the emerging economies of the Asia Pacific Economic Cooperation (APEC) forum.

Agreements between governments to recognize national conformity assessment mechanisms have a great potential to facilitate trade. A network of global mutual recognition agreements (MRAs) would enable manufacturers to test products once and obtain certification and acceptance in all national markets. At the regional level, for example, a successful conclusion to discussions within the

APEC forum on an MRA would provide significant new opportunities for U.S. trade expansion in rapidly growing markets of Asia.

- **RECOMMENDATION 5:** The Office of the U.S. Trade Representative (USTR) should continue ongoing mutual recognition agreement negotiations with the European Union. The USTR should also expand efforts to negotiate MRAs with other U.S. trading partners in markets and product sectors that represent significant U.S. export opportunities. Priority should be given to conclusion of MRAs on conformity assessment through the Asia Pacific Economic Cooperation forum.

As noted above, negotiations between the United States and the EU toward mutual recognition of conformity assessment mechanisms merit the continued high-level support of government, specifically the Office of the USTR. It is possible, however, that negotiations with Europe may not reach a timely or successful conclusion. Under these circumstances, failure by the Europeans to remove trade barriers in conformity assessment within a reasonable time period should lead to unilateral action by the United States, as authorized under our trade laws. Moreover, the USTR should use the full potential of targeted action on a unilateral basis under our laws, as appropriate, to remove barriers in other markets.

- **RECOMMENDATION 6:** The USTR should use its authority under Section 301 of the Trade Act of 1974 to self-initiate retaliatory actions against foreign trade practices involving discriminatory or unreasonable standards and conformity assessment criteria. In particular, if U.S.–EU negotiations do not succeed within two years in securing fair access for U.S. exporters to European conformity assessment mechanisms, the USTR should initiate retaliatory actions under Section 301.

Innovative export promotion programs, in combination with a systematic policy to lower trade barriers, have the potential for significant, long-term economic benefit. By providing technical assistance to countries in emerging markets as they construct modern standards and conformity assessment systems, the United States has a unique and valuable opportunity to facilitate future world trade.

- **RECOMMENDATION 7:** NIST should develop and fund a program to provide standards assistance in key emerging markets. The program should have four functions:

    (a) provide technical assistance, including training of host-country standards officials, in building institutional mechanisms to comply with the

Agreement on Technical Barriers to Trade under the Uruguay Round of the GATT;

(b) convey technical advice from U.S. industry, standards developers, testing and certification organizations, and government agencies to standards authorities in host countries;

(c) assist U.S. private-sector organizations in organizing special delegations to conduct technical assistance programs, such as seminars and workshops; and

(d) report to the export promotion agencies of the Department of Commerce (such as the U.S. and Foreign Commercial Service) and the USTR regarding standards and conformity assessment issues affecting U.S. exports.

ADDRESSING FUTURE CHALLENGES AND OPPORTUNITIES

The nation's ability to anticipate and respond to new developments in standards and conformity assessment will influence our future in many ways. There is the urgent need for increased federal data gathering and analysis on standards and conformity assessment. We require an ongoing capacity to analyze the economic effects of developments in domestic and international standards and conformity assessment systems. This new capacity would support improvements not only in our domestic systems, but also in our ability to monitor and anticipate international developments in key emerging areas such as environmental management standards.

In addition, wide dissemination of information to U.S. firms about standards and certification requirements in global markets is needed to improve prospects for future U.S. export expansion. Detailed and readily available information about international developments is especially important for our small and medium-size firms wishing to compete in global export markets.

• **RECOMMENDATION 8:** NIST should increase its resources for education and information dissemination to U.S. industry about standards and conformity assessment. NIST should develop programs focusing on product acceptance in domestic and foreign markets. These efforts should include both print and electronic information dissemination, as well as seminars, workshops, and other outreach efforts. Programs should be conducted by NIST staff or by private organizations with NIST cooperation and funding.

• **RECOMMENDATION 9:** NIST should establish a permanent analytical office with economics expertise to analyze emerging U.S. and international conformity assessment issues. The office should evaluate and quantify the cost to U.S. industry and consumers of duplicative conformity assessment requirements of federal, state, and local agencies. To support the work of the USTR and other federal agencies, including those involved in export promotion, it should also collect, analyze, and report data on the effects of foreign conformity assessment systems and regulations on U.S. trade.

• **RECOMMENDATION 10:** The USTR's post-Uruguay Round trade agenda, including work through the World Trade Organization, should include detailed analysis and monitoring of emerging environmental management system standards and their potential effects on U.S. exports. Technical assistance should be provided to USTR by NIST.

NOTE

1. For a comprehensive discussion of U.S. economic performance relative to other industrialized nations, see; the *Annual Report of the World Economic Forum*. Davos, Switzerland, 1994. Data series reported annually by the Bureau of Labor Statistics (BLS), U.S. Department of Labor on "International Comparisons of Manufacturing Productivity," and BLS data on relative levels of real gross domestic product (GDP) per employed person are relevant to cross-national comparisons of U.S. productivity and output. Numerous data sets which reveal relative competitive positions of the United States in service and manufacturing sectors are reported by the Organization for Economic Cooperation and Development and the World Bank in annual publications.

1

Introduction

M any facets of our daily lives depend on standards. Standards influence the products we use, the foods we eat, how we communicate, our means of travel, our modes of work and play, and many other activities. Standards may function to inform, to facilitate, to control, or to interconnect—frequently, a combination of such elements. They serve economic ends, enabling or catalyzing commercial transactions of all sorts. They also serve societal aims, such as protecting health, safety, and the environment.

There is no single, simple definition of *standard* that captures the broad range of meanings and uses of the term. There are, however, general characteristics of many or most standards that will serve as a working definition within this report. **A standard is a set of characteristics or quantities that describes features of a product, process, service, interface, or material.**[1] The description can take many forms, such as definition of terms; specification of design and construction; detailing of procedures; or performance criteria against which a product, process, etc., can be measured.

A standard can be formal or informal in varying degrees. Social customs—waving or shaking hands, for example—are informal standards. There is no codified, formal, "standard handshake" to which we refer for guidance when we meet someone; yet most Americans know and follow the standard.

A formal standard, by contrast, is one that has been formulated to meet a specific goal and written down by its developers so that others may use it. Formal standards can be written unilaterally by a designer, manufacturer, or purchaser. They can be developed through cooperation and consensus among a group of interested parties. Or they can be mandated by government. These paths often overlap, as when a group chooses to adopt a standard developed unilaterally by one of its members or when a government agency adopts a private standard by reference in a regulation or law. Mechanisms for U.S. and international standards development are considered in detail in Chapter 2.

This report is concerned primarily with formal standards. Formal standards impinge on our activities every day, with or without our conscious awareness. A ubiquitous element of American life, the automobile, offers many examples. We choose the grade of gasoline we put in our cars according to a formal standard for octane ratings. We use motor oil classified as SAE 10W-30, 10W-40, etc., against standards written by the Society of Automotive Engineers (SAE), a professional society that develops many standards for the automotive industry.[2] If we perform our own car repairs, we consult a manual of standards for parts and assembly published by the manufacturer.

Safety and environmental standards play a major part in our use of automobiles. We are protected by safety features that meet standards mandated by government or adopted voluntarily by automakers. In many states, cars must be tested and certified as meeting emission standards designed to protect air quality. To obtain a driver's license, we must pass a test that measures our skills and knowledge against formal standards, and when we drive, we obey standard traffic signals and laws.

These examples illustrate the influences of standards on the use of a product by its ultimate user, the consumer. The average car owner, however, may never think about an additional set of relevant standards—those that the automaker uses in designing and building the car. These standards play a role from the very beginning of the production cycle. Electronic data interchange standards enable teams of engineers to share designs on their computer workstations. Computer-controlled machine tools follow standard, coded instructions in cutting and welding sheets of metal—metal that meets material specifications for strength, rigidity, and other characteristics. The tools are calibrated to units of length, mass, pressure, and other quantities against references maintained by the U.S. Department of Commerce's National Institute of Standards and Technology (before 1987, the National Bureau of Standards) in Gaithersburg, Maryland.

In order to win the car manufacturer's business, suppliers of materials, parts, and services must meet performance specifications set by the manufacturer. The manufacturer also applies performance criteria to internal operations, as part of a continuous process of managing and improving the quality of his or her own products. Use of formal standards describing quality management systems, such as the International Organization for Standardization's ISO 9000 international standards series or the Department of Commerce's Malcolm Baldrige National Quality Award criteria, is increasing rapidly in this country, with consequences that are considered in this report.[3]

These examples demonstrate clearly that standards take many forms and serve many purposes. This report focuses primarily on formal standards for products and processes in manufacturing industries. This focus is determined by the committee's main objective, which is to examine the influence of domestic and international standards—through their development and use, as well as through testing and certification—on U.S. economic performance. Drawing upon this examination, a specific set of recommendations for improvements in public policy is outlined in Chapter 5.

FUNCTIONS OF STANDARDS

The first requirement for understanding the link between standards and economic performance is a systematic consideration of the functions of standards. Many schemes for classifying standards by function or goal have been developed.[4] The functions of standardization can be divided into seven categories (see Table 1-1). These categories are not mutually exclusive. Most standards serve more than one purpose.

Commercial Communication

Standards convey information about a product to the buyer in a consistent, understandable manner. This communication reduces the amount of work the buyer has to do to find out about the product's characteristics (or, alternatively, the work the seller must do to inform the buyer). In other words, standards reduce the transaction costs for buyer and seller.[5]

For example, the owner of a portable radio does not need to talk to a salesperson or experiment with various batteries in order to find ones that will work in the radio. He or she simply picks batteries labeled with the correct size—AAA, A, D, etc.—and makes the purchase, confident that they will fit the radio. If there were no standard, the consumer would have to invest time and effort in researching and trying out various batteries, or pay extra costs for expert advice and assistance. By contrast, in product sectors with less complete standardization—such as personal computer software—the task of identifying, buying, and installing compatible products can be time-consuming and expensive.[6]

The buyer–seller interaction is not a feature of the consumer marketplace alone. Procurement specifications set by purchasing departments of companies and government agencies are further examples of standards that convey information in support of a commercial transaction. Given the enormous range of transactions dependent to some degree on standards, standardization clearly facilitates commerce in the U.S. economy to a degree that is difficult to quantify, but unquestionably significant.[7]

Technology Diffusion

The previous discussion highlighted communication between a buyer and a seller. However, standards serve an additional communication-related function—recording technological advances in a form that others may reproduce and use. In

TABLE 1-1 — Functions of Product and Process Standards

CATEGORY	FOR EXAMPLE...
1. **COMMERCIAL COMMUNICATION** Standards convey information about a product to the buyer in a consistent, understandable manner.	(a) **construction materials**--standard dimensions, strengths, and durabilities make it easier for the builder to select materials for specific purposes (b) **film speed**--standard ratings (ISO 100, 200, 400, etc.) simplify matching film to photographic needs
2. **TECHNOLOGY DIFFUSION** A technological advance incorporated into a standard is more readily adapted and used by others.	(a) **personal computer architecture**--use of PCs expanded rapidly once IBM-compatibility standard came into being (b) **advanced materials (e.g., composites, ceramics)**--standards that describe processing and test methods allow duplication and improvement upon state of the art
3. **PRODUCTION EFFICIENCY** Standardization of parts, processes and products enables economies of scale in production.	(a) **automobile assembly line**--efficient mass production pioneered with the Ford Model T (b) **fast food chains (e.g., McDonald's)**--food, restaurant style, equipment, and procedures standardized for efficiency

4. **ENHANCED COMPETITION** When some or all of the features of different manufacturers' products conform to one standard, comparison is easier and competition sharper.	(a) **direct-dial long-distance telephone service**--competing carriers offer a standardized basic service; competition centers on price and extra services (b) **gasoline**--octane ratings allow consumer to compare similar products on the basis of price
5. **COMPATIBILITY** Standards defining interfaces enable products to work or communicate with each other.	(a) **Internet**--standard format for sending and receiving data enables communication among computers worldwide (b) **stereo system components**--various types of components can be connected with standard cables and jacks
6. **PROCESS MANAGEMENT** Manufacturers not only design products to conform to standards, they also organize the manufacturing process itself in accordance with standards.	(a) **numerically controlled machine tools**--standard computer languages allow rapid reconfiguration of production line (b) **quality assurance**--ISO 9000 series of standards guides firms in setting up and maintaining a quality assurance management system
7. **PUBLIC WELFARE** Standards are an important mechanism for promoting societal goals, such as protection of health, safety, and the environment.	(a) **health codes**--restaurants conform to sanitary standards that are backed up by inspections (b) **automobile air bags, seat restraints, and bumpers**--government-mandated crash protection

creating a new product or process, a designer may choose to use technological approaches already developed by others and incorporated into formal standards. This is an advantage if it enables the designer to avoid reinventing an already-existing innovation.

Standards play a key part in the process of technology diffusion. When a technological advance by a designer, researcher, or developer at one firm is incorporated into a standard and used by others, that advance is diffused throughout the industry. Innovations made at academic and government institutions can diffuse in the same way. This process raises productivity and industrial competitiveness. It increases efficiency by enabling firms to adopt standardized approaches, where appropriate, rather than reinventing technologies already developed elsewhere.[8] The adoption of best industrial practices and technologies throughout U.S. industry is a crucial factor supporting the nation's industrial competitiveness and economic performance.

If the standard incorporates a patented technology, the firm adopting it may have to pay a royalty (provided the owner is willing to license the technology). However, using the standard may still be more cost-effective than developing a unique approach. Choice between standardized and firm-unique technologies is one aspect of strategic standardization management, a managerial approach gaining adherents in some parts of U.S. industry.[9]

Productive Efficiency

A fundamental characteristic of twentieth-century manufacturing is mass production through interchangeable parts. First popularized by Eli Whitney in producing muskets for the Continental Army during the American Revolution, manufacturing standardization may have reached a peak in the 1920s at Ford's River Rouge automobile plant.[10] Standardization of parts and processes enables efficiency-increasing measures such as repetitive production, reduced inventories, and flexibility in substituting components on the assembly line. Production of standardized goods in great quantities for a uniform marketplace brings about significant economies of scale. These economies of scale benefit the producer through cost reductions, which may be passed on to the consumer in lower prices.[11]

Enhanced Competition

From the consumer's perspective, standards enhance the efficiency of purchasing in the commercial marketplace by placing products that conform to a standard in direct competition with one another. When some or all of the features of different manufacturers' products conform to one standard, the consumer's task of comparing is made easier—particularly with respect to price. These effects sharpen competition.[12]

It is important to note that standardization of products also limits the consumer's range of choices. Henry Ford's great River Rouge automobile plant was a model of productive efficiency through mass standardization. However, although this standardization lowered cost, it also reduced Ford customers' options to a single product: the black Model T.[13] Manufacturers compete on the basis of differentiated features and technological innovations as well as price. For this reason, there is the potential for excessive standardization to lead to decreased market choice and a reduced range of innovation. This may, in specific cases, outweigh the benefits of standardization.[14]

In short, both standardization and differentiation of products are necessary for society to gain the greatest economic and social benefit. The two are not necessarily incompatible. From the producer's perspective, standards for computerized design and machine-tool automation have fostered the spread of flexible manufacturing methods. These methods have made it increasingly possible for manufacturers to provide greater product variety without sacrificing economies of scale. In the late twentieth century, standardization of processes may be replacing standardization of final products as a key enabler of large-scale manufacturing efficiency.[15]

Compatibility

When products are used together, the standards defining their interfaces are important. For example, the widespread standard for stereo equipment interconnection makes it possible for a compact disk player made by one firm and an amplifier made by another to work together. The ability to mix and match components of a system—made possible by the interface standard—increases the consumer's choice and the breadth of competition in the market.

Compatibility standards—also known as interoperability standards—are especially important in industries that are organized into networks. An example is telecommunications. A telecommunication network gains value with increasing size, because each new member of the network adds to the scope of possible connections that each current member is able to make.[16] Standards describing how to attach to a network and use it are necessary for the network to form and grow. For example, formal definitions of formats for transporting electronic data through the global Internet make it possible for millions of people, using computers, to communicate with each other. It is the worldwide availability of Internet standards that makes the network's rapidly expanding size and scope possible.[17]

Process Management

Manufacturers not only design products to conform to standards, they also organize the manufacturing process itself in accordance with standards. Many of these are purely internal standards—the routines and procedures by which a company does its work. However, compliance with some external standards is a necessity for all firms. The U.S. Occupational Safety and Health Administration (OSHA) regulates many manufacturing processes to protect worker safety, while the U.S. Department of Agriculture (USDA) sets standards for the sanitation of meat processing facilities. Independent inspection and audit of production play a key part in the enforcement of process standards such as those set by OSHA and USDA.[18]

Firms may also perform—or hire independent auditors to perform—assessments of their procedures for ensuring product quality. This process-management approach to quality assurance, pioneered by W. Edwards Deming, is reflected in the internal quality programs of numerous manufacturers. It is also formalized in quality management system standards, such as the ISO 9000 series published by the International Organization for Standardization (ISO)—a private, international standards-developing organization in Geneva, Switzerland, with membership including national standards organizations from most countries of the world.[19] It is also reflected in the Department of Commerce's Malcolm Baldrige National Quality Award criteria. These standards furnish objective criteria for evaluating aspects of a firm's quality assurance processes. They also serve as one avenue for diffusion of best practices in quality assurance throughout industry as manufacturers invest the effort needed to adopt and conform to them.[20]

Public Welfare

Standards are an important means of promoting societal goals, such as protection of health, safety, and the environment. Government agencies at the national, regional, state, and local levels administer thousands of regulatory standards or technical regulations. These govern the characteristics of the products and services that manufacturers produce and the materials and processes that they use in producing them. Some regulatory standards are developed by government agencies, but many are developed within the private sector and adopted by government. OSHA and USDA guidelines noted above are examples of *process-oriented regulations*. The National Highway and Traffic Safety Administration's automobile bumper and air bag standards are examples of *product-oriented regulations*. Standards for treatment, storage, and distribution of drinking water have been developed by the private, non-profit NSF International and are used by the Environmental Protection Agency, state governments, and many public and private water utilities.[21]

In addition to federal standards, a range of regulations applies in the United States at the state, regional, and local levels. Building codes enforced by public inspectors set parameters for electrical wiring, plumbing, materials, and other aspects of construction. Many public jurisdictions apply water and air quality standards setting limits on pollutants and toxins that may be emitted into the environment. The automobile emission standards of the state of California, for example, are stricter than those applied at the national level.

CONFORMITY ASSESSMENT

Standards would be unable to fulfill any of the many purposes just outlined without some degree of confidence that manufacturers' claims for their products of conformity to standards are correct and justified. Such assurance can come from the firm's internal procedures for meeting standards; from review by an independent, private source outside the firm; from a government-mandated regulatory program; or from a combination of such elements.

Conformity assessment is the comprehensive term for procedures by which products and processes are evaluated and determined to conform to particular standards. As distinct from standards *development*, conformity assessment may be thought of as a central aspect of the *use* of standards. In the context of many commercial and regulatory uses of standards, measures to evaluate and ensure conformity are of as much or more significance than the standards themselves. They impose significant costs in manufacturing through testing, inspection, audit, and related procedures. The benefits that mitigate these costs accrue from the value added by increased buyer (or regulator) confidence that a product or service meets a standard.[22]

As noted previously, standards facilitate commerce by informing prospective buyers about aspects of products and services. An implicit element in this communication is the need for trust on the buyer's part that a product meets cited standards. Often, the reputation of the producer expressed in his or her brand name is adequate to establish this confidence. However, when a consumer looks for a recognized certificate of approval on a product, he or she gains additional assurance that the product has been independently tested and verified against applicable standards.[23] A well-known certification mark found on many products is the "UL" label. This mark is owned and managed by Underwriters Laboratories, a nonprofit institution that develops safety standards and tests and certifies many consumer and other products.[24]

The efficacy of regulation similarly depends on the reliability of measures to determine that products and processes comply with mandatory standards. Regulatory enforcement often—though not always—depends on assessment of the producer's compliance with regulations by an independent inspector or testing laboratory. Building inspection and water quality testing are examples of state and local government assessment of compliance with standards. Many independent, private-sector laboratories also test products for regulatory compliance.

Testing, manufacturer's declaration of conformity, independent certification, laboratory accreditation, and quality system registration are the key elements of the U.S. and international conformity assessment systems. Each of these elements figures prominently in the overall impact of standards on U.S.

economic performance. Chapter 3 contains a detailed examination of the organization, efficiency, and economic value of these processes.

STANDARDS, CONFORMITY ASSESSMENT, AND PUBLIC POLICY

The U.S. system for developing standards and assessing conformity to them is complex, decentralized, and multifaceted. It encompasses many kinds of standards, as the preceding discussion illustrates. It comprises many types of conformity assessment activities, conducted by manufacturers, independent private-sector testers and certifiers, and government regulators. Complexity is not necessarily evidence that the system cannot function well. Each aspect of the system has evolved to meet the conditions and requirements of a specific technology, industry sector, or public interest. Historical and political accident have also played their part in the system's development.

Assessing the U.S. System

How well does the U.S. system work? The purpose of this report is to analyze U.S. standards development and conformity assessment in the context of the nation's domestic and international economic performance. On the basis of this evaluation, improvements in U.S. public policy are recommended in three broad areas: (1) the *efficiency* of the domestic standards and conformity assessment system, including private-sector and government components; (2) the link between standards and U.S. *technological advance*; (3) the role of standards and conformity assessment in enhancing U.S. *export performance* and facilitating global trade.

Virtually every sphere of economic activity is influenced to some degree by standards and conformity assessment. This influence varies widely, however, among industry sectors. The functions of standards vary widely in their applicability to a given commercial activity. Compatibility standards, for example, are clearly much more important in network industries, such as telecommunications, than in industries producing stand-alone goods. Quality is important in all industries; however, setting standards for quality system management is a considerably more costly undertaking in industries with complex manufacturing or service processes than in those with simple procedures. Products that pose potential health, safety, or environmental risks are subject to many more regulatory standards than are relatively riskless products.

This report does not identify a discrete quantity representing the total economic impact of standards. The claim is justifiable, nevertheless, that standards have a highly significant role in the U.S. economy. The previously outlined categorization of standards by function (again, see Table 1-1) illustrates their ubiquity in our economy. The value of standards in the first category alone— informing buyers about products and services—is, if considered in the economy-wide aggregate, clearly significant. It is probably also impossible to quantify precisely. Such a task would require not only determining the number of commercial transactions that are influenced by one or more standards, but also estimating how efficiently those transactions would take place in the absence of standards—if they were to occur at all.

The number of standards and related conformity assessment programs (such as testing and certification) in the United States provides a rough sense of standardization's economic impact. Federal government standards—including regulations and public procurement specifications—total more than 50,000, of which Department of Defense standards are the majority. Private standards-developing organizations, professional societies, and industry associations account for more than 40,000 additional, formal standards.[25] These figures do not account for the much larger number of de facto industry standards. These are products, processes, and technologies that are established as standards not through formal procedures, but through widespread acceptance in the free market.[26]

How can the efficiency of the standards system be improved? Efficiency comprises two elements—cost of input and value of output. Chapters 2 and 3 analyze the costs, quality, processes, and organization of standards development and conformity assessment in the United States. The representation of U.S. interests in international standards-setting forums is considered as well. Although the scale of this report does not permit a detailed examination of individual industry sectors, principles are identified that apply across industries, as well as guidelines for linking more focused policy measures to appropriate industries.

The assessment of the U.S. system's efficiency in this report encompasses both its private- and its public-sector components. Many standards are developed entirely within the private sector and are applied voluntarily. This may occur through the formal procedures of consensus-seeking standards organizations or through the success of individual firms' products, services, or technologies in the competitive marketplace. Assessing the capacity of the U.S. system to generate appropriate standards in a timely manner is a complex issue, highly dependent on the specific industry and even the product or service in question. The vested interest in development and use of a standard—or choice among possible technologies to use in a particular standard—can vary widely among different producers, consumers, and government entities. This variation may, in some cases, strongly influence whether standards development activities produce the greatest possible public benefit with respect to such concerns as market efficiency and technological advance.[27]

Public Welfare

Government is both a major producer and a major user of standards. This report takes as given the fact that public needs may sometimes outweigh other concerns. There are, for example, health and safety concerns that justify imposition of regulatory standards despite the costs they impose on manufacturers and consumers. Government is also a major purchaser of goods and services, frequently by means of procurement standards and specifications. As a result, government agencies have a public interest in obtaining the best value for the public dollar through appropriate standards and efficient conformity assessment procedures.

Within these constraints, the standards system should work as efficiently as possible. It should be efficient both in terms of cost and in terms of enhancing, or not inhibiting, economic growth and technological innovation. Chapters 2 and 3 consider ways that improved coordination among standards developers and users in the public and private sectors can decrease costs and improve the functioning of the U.S. standards and conformity assessment system. Clearly, specific industry sectors vary in significant aspects. The objective of this report, however, is to identify the best opportunities for improving the overall system through public policy measures.

U.S. International Trade Policy

How can trade policy be improved to enhance export opportunities for U.S. firms and promote creation of jobs in the trade-oriented sector of the nation's economy? International trade is and will continue to be of growing importance to the U.S. economy. In the latter half of the 1980s, for example, exports accounted for 20 percent of the nation's total employment growth.[28] The continued increase in the share of U.S. economic performance dependent on international trade has raised the importance of international standards and conformity assessment issues to new heights. This report evaluates the performance of the U.S. standards and conformity assessment system not only as it meets the nation's domestic needs, but also with respect to facilitating global trade and U.S. exports.

Significant changes related to standards have taken place in the international trading system within the past year. The Uruguay Round of the General Agreement on Tariffs and Trade resulted in substantial revisions to the Technical Barriers to Trade provisions and creation of a new section on Sanitary and Phytosanitary Standards for food and agricultural products. Unfair trading practices based on standards, testing, and certification requirements are prohibited under these two sections of the agreement.[29]

The Uruguay Round provisions include updated and expanded coverage of governments' product approval regulations; inclusion of process and production method regulations; increased transparency to foreign firms of national and regional standards developing activities (such as those of the European Union); and promotion of mutual acceptance among trading partners of product test results, certifications, and other conformity assessment measures. In the context of the new World Trade Organization, progress continues in these and related areas. A current focal point of U.S. trade policy, discussed in Chapter 4, is the negotiation of agreements with our trading partners for mutual recognition of conformity assessment procedures. Chapter 4 assesses the potential utility and feasibility of such agreements for realizing the full benefit of trade enhancement provisions of the Uruguay Round, as well as forestalling the growth of trade barriers related to conformity assessment systems worldwide.

The emergence of significant new markets for U.S. products—particularly in Asia and Latin America—offers great potential for improvements in U.S. export performance. Many of the countries and regional trading groups that comprise these markets do not yet have well-developed systems for implementing standards or assessing conformity. In many of these markets, this condition is characteristic of the regulatory, public procurement, and private industry sectors alike. A key concern of this report is to examine how the United States can support efforts in developing markets to design and implement modern, open standards systems. Chapter 4 assesses the potential for enhancing U.S. exports and global trade through providing developing countries with appropriate technical assistance regarding implementation of standards and conformity assessment regimes.

NOTES

1. See Maureen Breitenberg, *The ABC's of Standards-Related Activities in the United States.* For additional examples, see ASTM, *ASTM and Voluntary Consensus Standards* (Philadelphia, PA.:ASTM, undated); and Carl Cargill, *Information Technology Standardization: Theory, Process, and Organizations.*

2. Maureen Breitenberg, *The ABC's of Certification Activities in the United States*, 5.

3. Maureen Breitenberg, *More Questions and Answers on the ISO 9000 Standard Series and Related Issues*, 1.

4. See Breitenberg, *The ABC's of Standards-Related Activities in the United States*, 3-5; Charles P. Kindleberger, *Standards as Public, Collective and Private Goods*, 1983, 378; and U.S Congress, (OTA), *Global Standards: Building Blocks for the Future*, 5-6.

5. Charles P. Kindleberger, *Standards as Public, Collective and Private Goods*, 384-385.

6. The enormous number of possible configurations of microcomputer hardware and software components produces strong incentives for both users and producers to reduce variety through standards. See Michael Hergert, *Technical Standards and Competition in the Microcomputer Industry.*

7. David Hemenway, *Industrywide Voluntary Product Standards*, 9-10.

8. For a discussion of business use of standards as a mechanism for adopting externally developed technologies, see Diego Betancourt, *Strategic Standardization Management: A Strategic Macroprocess Approach to the New Paradigm in the Competitive Business Use of Standardization.*

9. See American National Standards Institute (ANSI), *1993 Annual Report*, 8; and Betancourt, *Strategic Standardization Management.*

10. Breitenberg, *The ABC's of Standards-Related Activities in the United States*, 3; and OTA, *Global Standards*, 42.

11. See Hemenway, *Industry-Wide Voluntary Product Standards*, 21-26; and Kindleberger, *Standards as Public, Collective and Private Goods*, 384-388.

12. For a discussion of the effects of standardization on market efficiency under various conditions, see Farrell and Saloner, *Competition, Compatibility and Standards: the Economics of Horses, Penguins and Lemmings.*

13. OTA, *Global Standards*, 42.

14. Joseph Farrell and Garth Saloner, *Standardization and Variety*, 71.

15. A great deal of literature exists on the changing nature of manufacturing efficiency. See, for example, National Academy of Engineering, *Mastering a New Role: Shaping Technology Policy for National Economic Performance*, 28-39.

16. Paul A. David, *Some New Standards for the Economics of Standardization in the Information Age*, 217.

17. Computer Science and Telecommunications Board, National Research Council, *Realizing the Information Future: The Internet and Beyond*, 17-26.

18. For a partial listing of federal certification programs and their methodologies—including facility inspection—see Maureen Breitenberg, ed., *Directory of Federal Government Certification Programs.*

19. Breitenberg, ed., *Directory of International and Regional Organizations Conducting Standards-Related Activities*, 274-275.

20. Breitenberg, *More Questions and Answers on ISO 9000*; Curt W. Reimann and Harry S. Hertz, *The Malcolm Baldrige National Quality Award and ISO 9000 Registration*, 42-53.

21. NSF International Corporate Brochure, *Drinking Water Additives Program.*

22. See International Organization for Standardization (ISO), *Certification and Related Activities* for a detailed overview of types of conformity assessment activities and the costs and benefits, in qualitative terms, of each type.

23. ISO, *Certification and Related Activities*, 22.

24. Ross Cheit, *Setting Safety Standards: Regulation in the Public and Private Sectors*, 28; and Underwriters Laboratories (UL), *An Overview of Underwriters Laboratories*, brochure.

25. Robert B. Toth, ed., *Standards Activities of Organizations in the United States*, 4.

26. In the academic economics literature, there has been more analytical attention paid to free-market, de facto standardization processes than to coordinated, voluntary consensus processes. For an overview of the economics of de facto and consensus standard-setting, see Shane M. Greenstein, *Invisible Hands and Visible Advisors: An Economic Interpretation of Standardization*, 538-549.

27. For an overview of these effects, see Farrell and Saloner, *Competition, Compatibility and Standards.*

28. Office of the Chief Economist, Office of the U.S. Trade Representative, *U.S. Exports Create High-Wage Employment.*

29. Office of the U.S. Trade Representative, *Uruguay Round: Final Texts of the GATT Uruguay Round Agreements.*

2

Standards Development

S tandards serve many different purposes, as noted in the previous chapter. There are also many ways of developing standards. Figure 2-1 defines the three principal types of standards by development process. The first comprises consensus-building activities among private firms, technical experts, customers, and other interested parties. These groups write standards through a formal process of discussion, drafting, and review. Group members attempt to form consensus on the best technical specifications to meet customer, industry, and public needs. The resulting standards are published for voluntary use throughout industry. Standards arising from these processes are termed *voluntary consensus standards*. Examples range from dimensions of valve fittings in household plumbing to performance characteristics of automotive structural materials. A variety of private organizations produce voluntary consensus standards, including industry and trade associations; professional societies; nonprofit, standards-setting membership organizations, and industry consortia.

No single organization, public or private, controls the U.S. standards development system. The efforts of many U.S. voluntary consensus standards organizations, however, are coordinated by the private, nonprofit American National Standards Institute (ANSI). This organization sets guidelines for groups to follow in managing the consensus-seeking process in a fair and open manner. ANSI reviews and accredits many U.S. standards-setting organizations for compliance with these guidelines. It also approves many of the standards these organizations produce, designating them as American National Standards. These and other central roles that ANSI plays in the U.S. standards system, including representing

DE FACTO STANDARD	A standard arising from <u>uncoordinated</u> processes in the competitive marketplace. When a particular set of product or process specifications gains market share such that it acquires authority or influence, the set of specifications is then considered a de facto standard. *Example: IBM-compatible personal computer architecture*
VOLUNTARY CONSENSUS STANDARD	A standard arising from a formal, <u>coordinated</u> process in which key participants in a market seek consensus. Use of the resulting standard is voluntary. Key participants may include not only designers and producers, but also consumers, corporate and government purchasing officials, and regulatory authorities. *Example: photographic film speed--ISO 100, 200, 400, etc., set by International Organization for Standardization (ISO)*
MANDATORY STANDARD	A standard set by government. A procurement standard specifies requirements that must be met by suppliers to government. A regulatory standard may set safety, health, environmental, or related criteria. Voluntary standards developed for private use often become mandatory when referenced within government regulation or procurement. *Example: automobile crash protection—air bag and/or passive seat restraint mandated by National Highway and Traffic Safety Administration*

FIGURE 2-1   Types of standards.

U.S. positions in international standards organizations, are discussed in this chapter.

Not all private-sector standards are set through consensus. Many arise through competition in the commercial marketplace. When one firm's product becomes sufficiently widespread that its unique specifications guide the decisions and actions of other market participants, those specifications become a *de facto* market standard. De facto standards are sometimes called industry standards. A de facto standard is usually promoted by a firm or organization in pursuit of commercial benefits. These benefits include gaining economies of scale, protecting or increasing market share, and obtaining revenues from licensing of intellectual property, among others. The IBM personal computer architecture, established and promoted by IBM beginning in 1981, is an example of a de facto industry standard.[1]

De facto standards may arise without formal sponsorship, simply through widespread, common usage. Cultural norms and customs, including informal business practices, are unsponsored standards. The arrangement of keys on a typewriter or computer keyboard—the QWERTY keyboard, so named because of the placement of those letters in one row—is an example of an unsponsored, de facto technology standard. Although no firm or group of firms actively promotes the QWERTY standard, it remains the standard arrangement of most keyboards.[2] Most standards of interest in the context of this report, however, are actively sponsored by one or more organizations or individuals.

Mandatory standards are standards set by government with which compliance is required, either by regulation or in order to sell products or services to government agencies. Public-sector standardization encompasses many levels of government. Federal, state, regional, and local government agencies set regulatory standards on products and processes in order to protect health, safety, and the environment. They also produce specifications for public procurement of goods and services. Some of these standards are written by government agencies, whereas others are developed in the private sector and adopted by agencies. Even in the case of standards written by government, the process of development is not without private input or participation. For example, laws governing administrative processes—such as the Administrative Procedures Act—require public review and comment on proposed regulations. The *Federal Register* regularly publishes requests for comments on standards drafted by federal agencies. Technical requirements for safety devices on machinery, developed by the U.S. Department of Labor's Occupational Safety and Health Administration, are an example of mandatory standardization.

The boundary between voluntary and mandatory standards is not always distinct. Government standards writers frequently refer to privately developed, voluntary standards within the text of regulations and procurement specifications. Mandatory standards may cite voluntary standards in whole or in part, with or without additional criteria beyond those set in the referenced standard. For example, many of the regulations applied in state and local building codes require that electrical materials, such as wiring, meet portions of the National Electrical Code, a consensus standard developed by the private, nonprofit National Fire Prevention Association.[3] In addition, procurement specifications set by major manufacturers are, from the perspective of their suppliers, mandatory for doing business in the same way that government procurement standards are mandatory.

The mechanisms by which standards are developed are extremely diverse. There is no single process in the United States or worldwide for creating and adopting standards. There is great variability among different standards in such attributes as purpose, scope, specificity of requirements, relative technological sophistication, and speed of development. Many different types of organizations, companies, government agencies, and consumers are users of standards. The variables that affect the pattern of standards development in an industry or market sector include, among others, (1) industry size and concentration; (2) dominance of specific suppliers or buyers; (3) level and speed of technological advance; and (4) public interests such as safety, health, and environmental protection.[4]

Despite the diversity of U.S. standards development processes, however, some generalizations can be made that are useful in assessing the performance of the U.S. standards development system and providing guidance to policymakers. This chapter examines the major components of the U.S. system, turning first to the private-sector and then the public-sector components. Implications of the decentralized, market-oriented structure of U.S. voluntary standards development processes are highlighted, as well as interactions between voluntary and mandatory standards-setting mechanisms in the United States.

SCOPE OF THE U.S. SYSTEM

Standards exist for virtually all industries and product sectors. The 20 leading nongovernment standards developers in number of standards produced, for example, encompass a spectrum of industry sectors: aerospace; electronics; automotive and mechanical engineering; petroleum products; chemicals; pulp and paper; and cosmetics. This group also includes developers of safety-related standards such as those for fire protection, industrial hygiene, consumer product safety, and product testing.[5] Government standards at the federal, state, and local levels, including privately developed standards adopted by government, are similarly diverse. These encompass manufacturing, transportation, and communications equipment; environmental protection and public health; food, drugs, and consumer products; construction materials, such as electrical wiring, plumbing, wood, stone, and concrete; and the broad range of products procured for government use such as office equipment, vehicles, communications systems, and military hardware.[6]

The number of U.S. standards at a given point in time is difficult to identify. Table 2-1 details the approximate number of formal standards maintained in a current, active status by the main categories of public and private standards developers. The public sector list begins with the Department of Defense (DoD), which develops and maintains more formal standards than any other organization in the United States. The number of DoD standards was estimated by the National Institute of Standards and Technology (NIST, a branch of the Department of Commerce) at 38,000 in 1991. The number has begun to shrink, however, because DoD now decommissions more standards than it develops each year. Remaining federal procurement and regulatory standards bring the total of U.S. government standards to 52,000.[7]

The number of private-sector, voluntary consensus standards in the United States is 41,500.[8] Table 2-2 lists the 10 leading standards-developing organizations (SDOs) in the United States, by number of standards produced. There are three types of private standards-developing organizations. First are *technical and professional societies* that engage in technical standards development. These consist of organizations of individuals who practice a profession or discipline, frequently a branch of engineering. Second are *industry associations*, whose membership consists of firms in a specific industry or trade. The third group has the more generic designation of *standards-developing membership organizations*, whose membership is open to individuals interested in standardization. Unlike professional societies for which standards development is one among many functions, these organizations' primary focus is standards development and standards-related activities.

TABLE 2-1 — U.S. Standards, by Developer (active standards as of 1991)

FEDERAL GOVERNMENT	NUMBER OF STANDARDS
Department of Defense	38,000
General Services Administration (nondefense procurement)	6,000
Other federal (primarily regulatory)	8,500
Examples: Environmental Protection Agency, Occupational Safety and Health Administration, Federal Communications Commission	
Total	52,000
PRIVATE SECTOR [a]	**NUMBER OF STANDARDS**
Scientific and Professional Societies	13,000
Examples: American Society of Mechanical Engineers (ASME), Institute of Electrical and Electronics Engineers (IEEE)	
Trade Associations	14,500
Examples: National Electrical Manufacturers Association (NEMA), Computer and Business Equipment Manufacturers Association (CBEMA)	
Standards-Developing Membership Organizations	14,000
Examples: American Society for Testing and Materials (ASTM), National Fire Protection Association (NFPA)	
Total	41,500
Overall Total (Federal Government and Private Sector)	93,500

[a] Not including de facto industry standards.

SOURCE: Toth, Robert B., ed. *Standards Activities of Organizations in the United States*. NIST Special Publication 806. P. 4. National Institute of Standards and Technology. U.S. Department of Commerce. Washington, DC: U.S. Government Printing Office, 1991.

Data on numbers of standards must be treated with caution, for several reasons. First, the definition of what constitutes a standard is not exact. There may be uncertainty in whether to consider a product description, specification, definition of a term, or description of a procedure to be a standard. Different agencies and organizations may vary in their interpretation of such cases. The context of the discussion is also significant. Given this report's primary focus on the links among standards, conformity assessment, and domestic and interna-

TABLE 2-2 — Top 10 Private Standards-Developing Organizations (active standards as of 1991)

TOP 10 DEVELOPING ORGANIZATIONS IN THE U.S.	NUMBER OF STANDARDS
American Society for Testing and Materials	8,500
Society of Automotive Engineers	5,100
U.S. Pharmacopeia	4,450
Aerospace Industries Association	3,000
Association of Official Analytical Chemists	1,900
Association of American Railroads	1,350
American Association of State Highway & Transportation Officials	1,100
American Petroleum Institute	880
Cosmetic, Toiletry & Fragrance Association	800
American Society of Mechanical Engineers	745

tional performance of U.S. manufacturers, the discussion here is limited to product and process standards. Professional certification, for example, such as that required of accountants, lawyers, and health care providers, is a type of standard that is not considered in this report.

Second, the distinction between a single standard with many sections and a series of separate, but related, standards may be arbitrary. The American Society of Mechanical Engineers (ASME) produced and regularly updates the Boiler and Pressure Vessel Code, a single standard running into thousands of pages. The code currently has 11 major sections covering design, fabrication, inspection, and safe operation of boilers, pressure vessels, and nuclear power plant components.[9] The choice of writing a set of specifications as a single standard or a series of standards is made by each developing organization or agency, according to its own guidelines.

Third, not all published standards have equal influence in the economy. Some voluntary standards fail to achieve widespread acceptance or use in the marketplace. The most widely used 15 to 20 percent of standards developed by private organizations accounts for 80 percent of those organizations' orders for printed copies of standards.[10] These standards may be presumed to have greater economic and technological significance than those that are rarely used. A standard applied at a critical point in a system, market, or industry, however, could have an impact far greater than the number of copies ordered from its publisher would indicate.

Both voluntary and mandatory standards may become technologically obsolete, yet remain in a technically active status. For example, an organization may choose to maintain an outdated standard for the benefit of persons who own or acquire a piece of old equipment and need access to technical information in order to operate or maintain it. Alternatively, a standard may remain in existence simply for lack of incentive to unlist it. As many as 25 to 30 percent of U.S. government and private standards have been estimated to be obsolete.[11]

Finally, two groups of product and process standards are omitted from the data in Figure 2-1. First, information on numbers of state and local government standards is extremely limited and fragmentary. These standards are concentrated in such areas as building and construction materials, workplace safety, environmental protection, agriculture, and foods.[12] Second, de facto standards are also excluded from the table. The same difficulties in enumerating public and private formal standards apply in the case of de facto standards set by firms through market competition. In addition, the distinction between a product that sets a standard, influencing the design of others, and a product that is simply one among many is highly subjective. The absence of a formal, institutional process for designating de facto standards compounds the difficulty of identifying, much less quantifying, the output of de facto standards development efforts.

These factors, among others, make it clear that neither a determination of the economic impact of standards activities in the United States nor an overall assessment of the U.S. system should focus closely on the quantities of standards produced. Valid assessments depend, instead, on examination of the efficiency and effectiveness of standards development in relation to the needs of industry, government, and society; the economic and technological implications of the U.S. system's characteristics; and the efficacy of existing mechanisms for strengthening and improving the system.

PRIVATE-SECTOR STANDARDS

Efforts to coordinate standards development in the United States began to develop momentum early in this century. One factor spurring these efforts was the realization that technical standards were needed to ensure the safety of many new products of the industrial age. The first version of the American Society of Mechanical Engineers Boiler Code was written in 1914, in response to serious hazards posed by poor-quality boilers, which were prone to explode. The code—today, the Boiler and Pressure Vessel Code—performs several of the functions of standards outlined in the previous chapter. Most significant is its role in protecting safety by providing a standard against which unsafe boilers, components, and manufacturing methods can be identified and rejected. Large portions of the code have become mandatory through reference in government regulation in the United States and many other countries.[13] The code also acts as a guide for manufacturers in the techniques of producing and maintaining safe boilers, pressure vessels, and nuclear reactors. In this way, the code fosters the diffusion of best practices throughout the industry.[14]

A large fire in downtown Baltimore in 1904 was another impetus to standardization. Fire engines from other cities that came to assist the Baltimore fire department were unable to connect their hoses to the local hydrants. The disaster, which included the loss of 1,526 buildings, could have been prevented if hose connections had been standardized, as they are today.[15]

Safety concerns were not the only factor that fostered industrial standardization in the United States. The economies of scale afforded by mass production are driven by standardization of parts and processes, as discussed in Chapter 1. The automotive industry was an early proponent of standardization, not only within each manufacturer's own plants, but industry wide. There were several incentives for standardization across the industry: It enabled parts suppliers to produce large quantities for multiple customers, such that suppliers could gain economies of scale and lower their costs. Suppliers passed these savings on as lower prices to automobile manufacturers. In addition, standardization meant that if one supplier went out of business (a frequent occurrence in the early years of the industry), shortfalls of parts could be made up by other suppliers without a delay for reconfiguring their machinery to new specifications. Standards also allowed manufacturers to impose minimum quality criteria on their suppliers, particularly for steel. In general, standardization benefited both suppliers and manufacturers throughout the industry.[16]

Coordinating standards development among different automotive firms became the responsibility of the Society of Automotive Engineers (SAE). SAE was (and remains) a professional society whose membership spanned the industry, including both manufacturers and suppliers; was independent of any one firm or set of interests; and had the technical competence for the required work. Its success in reducing the variety of parts and in promoting interchangeability and quality was such that the National Automobile Chamber of Commerce, an industry trade association, estimated in 1916 that SAE standards yielded cost reductions of 30 percent in ball bearings and electrical equipment and 20 percent in steel.[17]

Economic Rationale for Consensus Standardization

These examples from the history of standardization illustrate one of the most important economic aspects of standards. Uncoordinated market mechanisms alone do not ensure that necessary standards are set. Firms acting in isolation are not as effective at setting an industry standard as producers, customers, and other interested parties acting in coordination. Even in situations in which all participants in an industry sector would benefit from standardization, cooperation and communication among them are usually necessary for a standard to emerge.[18]

A primary reason cooperation is necessary is that standardization requires gathering information and developing compromises among the needs, interests, and capabilities of many different interested parties. It is not impossible for a single firm to accomplish this task by marketing a product that meets the needs of diverse parties. In some cases, there are economic incentives to make the attempt. If a firm is successful in promoting its own, proprietary solution to a technological need and sets an industry-wide de facto standard, it may reap large benefits from dominating the resulting market. Microsoft Corporation's MS-DOS operating system is an example of success in this area. However, the costs of coordinating and accommodating multiple interests can be high. In addition, if a rival firm simultaneously attempts to set a competing standard, the companies can become caught in a winner-take-all game of investment and price cutting. In this case the benefits of success are outweighed by the costs of competing to set the standard.[19]

Another important reason the uncoordinated market can sometimes produce too few standards derives from the public nature of standards. When a standard has been set, everyone may use and benefit from it. This is true whether or not they participated in its development. The potential thus exists for free riders to benefit from standards-setting work done by others. In economics terminology, standardization is a public good.[20] A standard can be used any number of times without depleting its utility. The more widely a standard is used, in fact, the more valuable it becomes—not only to those who originally developed it, but to all users. Communication, compatibility, economies of scale, and other benefits of standards all increase as those standards become more widespread. Conversely, if a standard is little used, its value is limited. Although coordination among participants in an industry takes time and effort, it increases the likelihood that the standard will become widely used and thus acquire value.

These theoretical examples are not meant to show that the free market is unable to produce standards. They demonstrate that individual firms acting alone may be unable to justify the cost of developing and promoting their own proprietary standards against the risk that their efforts will fail to establish a de facto standard. A firm that bears the costs of developing a standard by itself cannot generally capture rewards equal to the overall social and economic benefit that accrues from standardization. (An exception is noted in the next paragraph.) As a result, market incentives alone are not sufficient to encourage firms, acting in isolation, to produce as great a degree of standardization as would be most economically beneficial to the industry or to society at large.[21]

An exception, applicable particularly in the communications and information technology industries, is the establishment of *network compatibility standards*. A detailed body of recent economics literature examines the incentives facing firms to establish compatibility standards, such as telecommunication system interfaces and computer operating systems.[22] These standards have unique economic properties, because they exhibit unusually strong returns to scale. Specifically, the more widespread a given network standard becomes, the greater does the incentive become for additional users to adopt that standard rather than be left as "orphans," incompatible with other systems. A firm that builds momentum behind its standard can benefit from a bandwagon effect in which users rush to adopt that standard. IBM Corporation in the 1960s, with System 360, and Microsoft Corporation's MS-DOS operating system in the 1980s are familiar examples of proprietary, de facto standards that conferred enormous economic benefits on their sponsors as computer users adopted them.[23] Responses of consensus standards-developing organizations to the challenges posed by compatibility standards are discussed later in this chapter.

In most cases, as noted above, uncoordinated market competition among firms promoting their own, proprietary solutions to particular market needs will not lead to as much standardization as is theoretically desirable for the economy as a whole. *Voluntary consensus standards* are an effective, rational response to this economic dilemma of standardization in the free market. As the example of the early automobile industry and the Society of Automotive Engineers illustrates, industry participants working together can share the effort of developing standards and gain mutual benefits from the results. Most of the formal standards used in the U.S. private sector are developed by private standards-developing organizations, such as SAE. These organizations have developed procedures to foster communication, coordination, and consensus in order to overcome the limitations of the uncoordinated marketplace and achieve industry-wide standardization.

Many foreign countries, including key U.S. trading partners in Europe and Asia, have a central, primary national standards-developing body. This is usually a government-chartered private organization or a quasi-public agency, rather than a direct agency of the government. Examples include Germany's Deutches Institut fur Normung (DIN), the British Standards Institute (BSI), and France's Association Francaise de Normalisation (AFNOR).[24] It is important to note, however, that even in countries in which a government agency sets national industrial standards, private-sector input plays a vital, pervasive role. It is impossible, given the breadth of technical and commercial expertise required to write standards, for all industry sectors to reside in any one organization. The resources of a national standards organization must always be supplemented with private-sector manpower, technical knowledge, and understanding of marketplace needs in order to develop useful standards. At the level of the technical committees—the volunteers who write the standards—the differences between foreign, relatively centralized standards systems and the U.S. system are negligible.[25]

It is important to note that issues related to ownership of private standards can influence the role they play in particular markets. Many standards developers, for example, offset expenses and generate income through sales of standards documents, to which they hold the copyright. For many SDOs, publishing is a significant source of operating revenue.[26] In addition, license fees and royalties are often paid to owners of patented innovations incorporated into standards. These fees can be a significant incentive for firms to innovate and develop new technologies and to permit their incorporation into standards. As previously noted, some firms gain substantial benefits from owning the rights to a proprietary technology that becomes a de facto industry standard. Prohibitively high payments to the owner of technology in a given standard, however, will deter its spread and motivate other parties to develop an alternative standard.[27] In most consensus standards organizations, owners of intellectual property incorporated into a formal standard agree to license proprietary technology at reasonable terms.[28]

Voluntary Consensus Standardization Processes

In comparison to most foreign systems, the institutional structure of the U.S. voluntary consensus standards system is highly decentralized. The United States has more than 400 private standards developers. Most are organized around a given industry, profession, or academic discipline. About 275 engage in ongoing standards-setting activities. The remainder have developed standards in the past—usually few in number—and occasionally update them.[29] There are three main types of U.S. standards-developing organizations: professional and technical societies, industry associations, and standards-developing membership organizations, discussed later in this chapter.[30]

All standards-developing organizations, to varying degrees, seek to overcome economic obstacles to standardization. The typical method for achieving this goal is to coordinate participation of volunteer technical experts in standards-writing committees. Each technical committee is responsible for standards in a particular area of product, process, or technology, although overlap does sometimes exist among different committees' scope of work. Committee membership is generally selected to represent a diversity of interests and viewpoints. Committees—or, in some cases, working groups that are subsets of a committee—meet on a semiregular basis over a period ranging from weeks to years. The first step in developing a standard is to identify an area of marketplace need requiring a standardized technical solution. Once a scope of work is set, draft technical standards are proposed, discussed, revised, and voted on. Consensus is, in most organizations, a key goal. Although negative votes do not prevent a standard's adoption, they must generally be considered and responded to in writing.[31]

Participants in a technical committee may propose, as foundations for a standard, technologies developed by their respective firms. Success in this effort may yield a marketing advantage or a technological head start over other companies whose technologies are not chosen. Alternatively, the committee may develop a compromise standard incorporating aspects of multiple proposals.[32]

After review, comment, and approval by the SDO's oversight board and membership at large, the organization publishes the standard. If the organization uses ANSI-accredited procedures, it may choose to have the standard approved and distributed by ANSI as an American National Standard. ANSI does not review the standard for technical merit but, rather, certifies that it was developed through open, consensus-oriented procedures and does not unduly duplicate or conflict with existing standards. The standard's usefulness to interested parties in the relevant market sector—manufacturers, purchasers, regulators, testing laboratories, certifiers, and others—largely determines whether it gains widespread acceptance. A technologically obsolete, commercially nonviable, or otherwise unsatisfactory standard will be neglected. Such a standard will eventually be discontinued by the SDO. Broad dissemination and use of the standard, however, are presumably in the interest of those who participated in writing and approving it. These individuals and the firms or associations they represent are therefore likely to use and promote the standard.

There is ample opportunity for U.S. industry to participate in voluntary consensus standards development and ensure that it meets U.S. economic needs. Both manufacturers and their customers take part in standards setting through industry associations such as the Computer and Business Equipment Manufacturers Association (CBEMA); the Gas Appliances Manufacturers Association; and the Alliance for Telecommunications Industry Solutions, consisting of telecommunications service and equipment companies. Firms also pay salary and travel expenses for employees who serve as individuals in the work of professional societies and standards-developing membership organizations such as SAE, the American Society for Testing and Materials (ASTM), and the Institute of Electrical and Electronics Engineers (IEEE). Small firms may not have sufficient resources to devote personnel to technical committees. They can, however, monitor and submit technical inputs to the process through industry and trade associations. Firms are often active in developing standards within all three types of organizations and through de facto marketplace competition at the same time.

Standards-developing organizations vary widely in size, number of standards produced, breadth of industries and technologies covered, profile of membership, and geographic scope, among other factors. Nearly all, nevertheless, share two important features.[33] First, they operate on the basis of *consensus*. Simple majority vote among participants in a standards-writing project is almost never sufficient to establish a standard. The consensus principle makes good sense in the context of the standards developer's mission. To produce standards that will achieve economies of scale, consumer safety, advancement of technology, compatibility, and other benefits of standardization, the standards must be accepted and used by as many firms and individuals as possible. Unless the standard is subsequently mandated as part of a government regulation or procurement specification, its acceptance by potential users is voluntary. Standards adopted as mandatory by government, moreover, are usually more effective if they reflect consensus among affected parties. A consensus among interested parties during the design of a standard clearly increases its prospects for broad acceptability.

The second feature common to most standards-developing organizations is administrative *due process*. These groups have formal policies governing such facets of standards development as technical committee membership; setting the scope of proposed standards; drafting and revising standards; voting within committees; review of draft standards by higher authority within the SDO; and balloting and approval by the membership at large.[34] Due process in SDOs bears many resemblances to public administrative procedures law. Laws governing public agency decision-making processes have such aims as representation of multiple interests; objectivity and fairness of procedures; public access to information about agency actions; and accountability of the agency through formal appeals. Analogous features—public notice and comment, appeals, multiple interest group representation, and democratic procedures—are all to be found in the policies of most formal standards-developing organizations as well.[35] These procedures increase the likelihood that a technical committee will reach a broad-based consensus, enhancing the value of the resulting standard.

Formal procedures, such as open participation and review, also serve as protection against allegations of collusive behavior for participants from competing firms. Consensus standards development is, in fact, well tolerated by U.S. antitrust law and precedent.[36] There have been few successful antitrust lawsuits related to U.S. voluntary product standards. In each case where the suit was successful, it was the subsequent interpretation of the standard by some other party, such as a certifier, that was deemed anticompetitive. One example is *American Society of Mechanical Engineers* v. *Hydrolevel Corporation*, a 1982 case in which a standards developer, ASME, was defeated in an antitrust suit. It was the actions of a committee interpreting product compliance with the ASME Boiler Code that was found to be anticompetitive—not the code itself or the process by which it was written.[37]

The principles underlying consensus standards development evolved over a period of many years, within many different SDOs. Each organization applies the principles in different ways, with procedures and objectives specific to the needs of its industry sector or professional competence. Authority in the U.S. standards-developing system, consequently, is highly decentralized and linked to specific industry sectors. Adherence to the basic principles, however, is actively promoted through the central, coordinating function of the American National Standards Institute. ANSI is not a standards developer but, rather, a nonprofit organization that coordinates and supports the U.S. consensus standards development system. U.S. standards developers desiring ANSI accreditation of their procedures and standards must follow ANSI guidelines for consensus, open participation, and due process. Through accreditation, ANSI seeks to promote and perpetuate core principles of the U.S. voluntary standards system.[38] Other ANSI functions in the U.S. system are discussed later in this chapter.

Standards-Developing Organizations

As previously noted, there are more than 400 standards developers in the U.S. private sector. Most of these belong to one of three main categories: industry associations, professional societies, and standards-developing membership organizations. In addition, consortia are playing an increasingly important standards development role, particularly in industries characterized by rapid advance of technology. Testing and certification organizations, such as Underwriters Laboratories, NSF International, and the American Gas Association, are discussed in Chapter 3. They represent a specialized category of organization responsible both for developing standards in certain sectors—typically related to health and safety—and for providing associated testing and certification services.[39]

Professional Societies Professional societies are individual membership organizations that support the practice and advancement of a particular profession. Several such societies, particularly in the engineering disciplines, develop technical standards. The goal of these SDOs is generally to find the best technical solution to meet an identified need. Participants in standards committees serve as individual professionals, not as representatives of the firm they work for. If more than one employee of a single firm serves in a committee, each still has a full vote in committee deliberations. Marketing considerations, however—such as securing commercial advantage for participants' firms—are in many cases secondary to technical factors in committee deliberations.[40] Funding for these SDOs is principally from publication and sales of standards, as well as direct services to industry.

> Institute of Electrical and Electronics Engineers Standards. IEEE Standards is a division of IEEE, Inc., an engineering professional society, founding member of ANSI, and ANSI-accredited standards-developing organization. The IEEE has a membership of more than 300,000 engineering professionals worldwide. IEEE Standards publishes more than 600 standards. Its area of expertise is electrotechnology, which ranges from electrical circuitry to artificial intelligence to aerospace. A Standards Board composed of voluntary industry and government representatives and 10 committees review requests from technical groups to initiate standards projects. After a consensus process, standards are approved by the Board and published as IEEE standards. IEEE participates in the United States National Committee (USNC) of the International Electrotechnical Commission (IEC), and the U.S. Technical Advisory Group (TAG) to the ISO and IEC Joint Technical Committee on Information Technology (JTC1).[41]

American Society of Mechanical Engineers. A founding member of ANSI, ASME is a professional society with an international membership of more than 100,000. It publishes 745 standards. In addition to standards development, ASME is involved in publishing, technical conferences and exhibits, engineering education, government relations, and public education. The ASME Council on Codes and Standards oversees 10 boards that supervise more then 700 committees. Drafts of standards are approved by committees and opened up to public comment after which, if necessary, they are redrafted and published as ASME standards. ASME is well known for its Boiler Code first published in 1914, and most recently revised and published as the Boiler and Pressure Vessel Code in 1992.[42]

Industry Associations Industry associations, also known as trade associations, are organizations of manufacturers, suppliers, customers, service providers, and other firms active in a given industry sector. Their mission is to further the interests of their industry sector, including the development of technical standards. Many industry associations develop standards or sponsor their development through a subsidiary or associated SDO. Funding is primarily through members' dues. Members of technical committees typically serve as representatives of their firm. Each firm carries equal weight in committee voting, regardless of the number of experts it sends to participate in the committee's standards development work. Industry association SDOs are likely to be more openly responsive to commercial market concerns in their technical decisionmaking than other types of SDOs.

Computer and Business Equipment Manufacturers Association. Accredited Standards Committee X3, Information Processing Systems (ASC X3). Created in 1916, the CBEMA is a professional trade association involved in the information processing, communications, and business products industry sectors. It maintains an ANSI-approved voluntary consensus program and sponsors the ASC X3 Secretariat, which oversees the legal, financial, and procedural work of the committee. The ASC X3 has 41 members including producers and consumers in the information technology industry, and it manages more than 500 projects. ASC X3's Operational Management Committee (formerly, the Standards Planning and Requirements Committee, SPARC) reviews submitted standards proposals and reports on its activities to the ASC X3. A proposal, after receiving X3 approval, is assigned to a technical committee for development into a standard. To complete the consensus process, the draft standard is redrafted, voted on again, and then sent to ANSI for final approval. The Strategic Planning Committee (SPC) manages the standards process and helps to define the role of information technology standards in the industry.[43]

Association for the Advancement of Medical Instrumentation (AAMI). AAMI is a voluntary membership organization of about 5,000 health care professionals from industry, health care facilities, academia, research centers, and government agencies such as the Food and Drug Administration (FDA).[44] AAMI works to advance patient care and medical technology through certification of biomedical engineers, education, technical publications, and development of medical device standards. AAMI is active in the development of international standards through sponsorship and representation on ISO and IEC committees and U.S. and ISO technical advisory groups (TAGs). It also maintains relationships with European counterpart organizations involved in standards development.[45] In 1994, AAMI became the international secretariat for a new ISO committee developing standards for quality management systems in the health care equipment industry.

Membership Organizations   Unlike industry associations and professional societies, standards-developing membership organizations have standards development as their central activity and mission. They do not limit their membership to an industry or profession, and they tend to have the most diverse membership among all SDOs. Their procedures tend to have the strictest due process requirements. Publishing and selling standards documentation accounts for the majority of their revenues. Membership fees are generally relatively low, facilitating participation by individuals not sponsored by an employer.

American Society for Testing and Materials. Established in 1898, ASTM is one of the world's largest voluntary consensus SDOs. It has an international membership of more than 35,000 and maintains 132 technical standards-writing committees. ASTM publishes standard test methods, specifications, practices, guides, classifications, and terminology for materials, products, systems, and services. Approximately 33 percent of ASTM's sales of standards are to international users, and many of its standards become de facto international standards. It produces an annual 70-volume set of more than 9,000 ASTM standards that are used worldwide.[46]

National Fire Protection Association (NFPA). The NFPA was established in 1896 in response to the need for uniform installation of sprinkler systems. NFPA publishes 280 specific fire safety standards.[47] NFPA is an independent, nonprofit, voluntary membership organization with an international membership of more than 60,000 individuals and 115 national trade and professional societies. It maintains some 235 consensus standards committees. After public comments are reviewed and committees reach consensus, the NFPA membership votes as a whole on adoption of standards as voluntary national standards. NFPA is also actively involved in public fire protection, fire analysis and research, government relations, and public education. Its standards are used in the fields of aviation, chemicals, engineering, hazardous materials, health care, marine fire protection, and signaling systems, among others. It also publishes the National Fire Codes and National Electrical Code, which are referenced in many state and local building regulations.[48]

Consortia Standards consortia are a response to the rate of technological advance outpacing consensus standards development in some industry sectors.[49] They focus particularly on compatibility standards. Examples include the Corporation for Open Systems (COS) and the Manufacturing Automation Protocol (MAP) user group. COS, a vendor consortium, was established to promote the Open Systems Interconnection (OSI) suite of computer interconnection protocols. Currently, COS is active in testing OSI products for conformance to the standards. By contrast, MAP is a user consortium, created to pressure vendors of manufacturing automation systems to develop compatible products.

Participation in standards-setting is generally limited to consortium members. Requirements for openness, consensus, and due process are less strict than in other standards-developing organizations, primarily to speed the development process. In fact, standards produced by consortia represent a hybrid stage between de facto industry standards and full consensus standards. To gain acceptance of their standards in the marketplace, consortia may seek after-the-fact accreditation of the standards through ANSI procedures. This is particularly the case, for example, for consortia wishing to promote international acceptance of their standards through ISO and IEC.

American National Standards Institute

The American Engineering Standards Committee was formed in 1918 as a federation of several prominent SDOs. In the 1960s, after several name changes, it became the American National Standards Institute (ANSI). Its principal missions are to coordinate and strengthen the U.S. voluntary consensus standards development system; to promote awareness and use of voluntary standards; and to represent U.S. interests in international standardization bodies.[50] ANSI is a nonprofit organization with annual revenues in 1993 of $16.7 million.[51] ANSI membership includes approximately 1,300 companies; 35 government agencies; and more than 260 technical, trade, labor and consumer groups.[52]

ANSI's organizational structure is decentralized (see Figure 2-2). ANSI's intent is for standards developers and users in different industry and technology sectors to be able to manage the development of standards at the level and direction appropriate for each sector. For example, ANSI members in the information technology industry emphasize international standardization, whereas consumer and workplace safety and health standards are developed with a focus on U.S. national standards. Four member councils discuss issues from their constituents' perspectives, bringing them before the ANSI Board of Directors as needed. These are the Company, Consumer Interest, Government, and Organization Member Councils.[53]

As noted earlier, one of ANSI's key means of carrying out its mission is to accredit U.S. standards developers. ANSI accredits both the organizations that develop standards and the standards themselves. (See Figure 2-3 for an overview of ANSI-accredited standards processes.) Accreditation is based not on the technical merits of standards but, rather, on the procedures used to develop them. Adherence to open participation, due process, and consensus procedures is necessary for an SDO to become an ANSI-accredited standards developer. ANSI accepts three different types of standards developers. Accredited organizations include most of the largest U.S. SDOs. Accredited standards committees (ASCs) write standards for a specific industry or technology sector, with administrative support provided by an interested host organization. An example is ASC X3 for Information Processing, whose secretariat is provided by CBEMA. Accredited sponsors are smaller groups that seek comment on and approval of their standards through a ballot of interested parties. These groups are usually formed to write one or a select few standards for a narrowly focused application.[54]

ANSI-accredited organizations may submit standards they have developed for ANSI approval as American National Standards. ANSI publishes American National Standards developed by some, mainly smaller, standards developers. Larger SDOs, such as ASTM, ASME, IEEE, and NFPA, publish standards under their own organizational name, even if they have been accredited as American National Standards. For example, more than half of the standards listed in the IEEE Standards Catalog are indicated, by footnote, as "recognized by the American National Standards Institute."[55] The National Electrical Code, published by NFPA, is approved and identified by ANSI with the designation ANSI/NFPA 70; NFPA's own publications, however, refer to it simply as NFPA 70.[56]

Although ANSI is not a standards developer, as noted above, it publishes American National Standards developed by some of the groups it accredits. This activity has been a source of conflict between ANSI and some of the larger SDOs. Approximately 65 percent of ANSI's $16.7 million gross income (based on 1993 data) is generated from sales of standards and other publications. Net income from publication sales provides for 34 percent of ANSI's core (nonpublishing) expenses, which are not fully funded by membership dues.[57] The SDOs' main objection is to ANSI's accrediting and providing publication services to smaller trade and professional associations to produce standards by the canvass method, rather than through a committee process.[58] (Some standards developers rely on both canvass and committee methods at different times.) These groups can

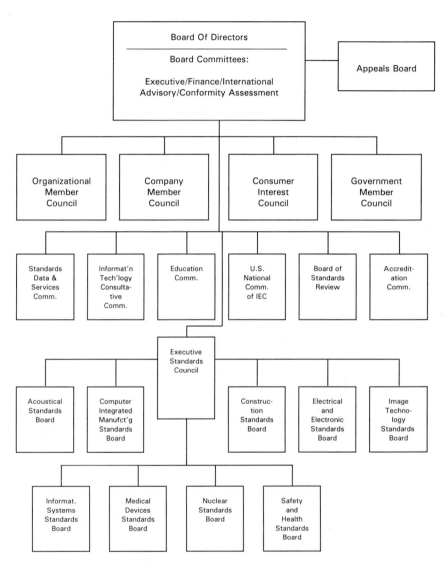

SOURCE: American National Standards Institute. *The U.S. Voluntary Standardization System: Meeting the Global Challenge*. Second edition. Reprinted with permission. New York: ANSI, 1993.

FIGURE 2-2   American National Standards Institute: Organization.

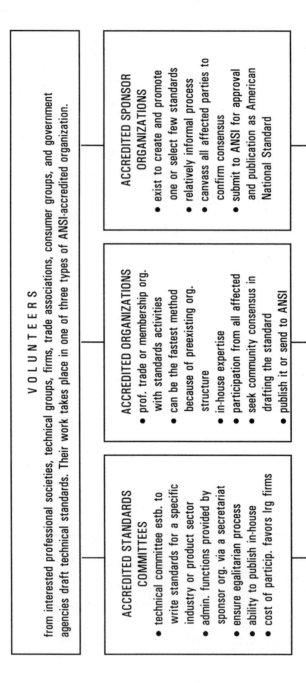

VOLUNTEERS

from interested professional societies, technical groups, firms, trade associations, consumer groups, and government agencies draft technical standards. Their work takes place in one of three types of ANSI-accredited organization.

ACCREDITED STANDARDS COMMITTEES

- technical committee estb. to write standards for a specific industry or product sector
- admin. functions provided by sponsor org. via a secretariat
- ensure egalitarian process
- ability to publish in-house
- cost of particip. favors lrg firms

ACCREDITED ORGANIZATIONS

- prof. trade or membership org. with standards activities
- can be the fastest method because of preexisting org. structure
- in-house expertise
- participation from all affected
- seek community consensus in drafting the standard
- publish it or send to ANSI

ACCREDITED SPONSOR ORGANIZATIONS

- exist to create and promote one or select few standards
- relatively informal process
- canvass all affected parties to confirm consensus
- submit to ANSI for approval and publication as American National Standard

A standard can be written/drafted in final committee format within 18 months, or the process can take as long as 5 years. Timing depends on interest in the standard, its technical complexity, and the makeup of the committee. Secretariats or boards of directors provide oversight and act as liaison to ANSI and other standards development organizations.

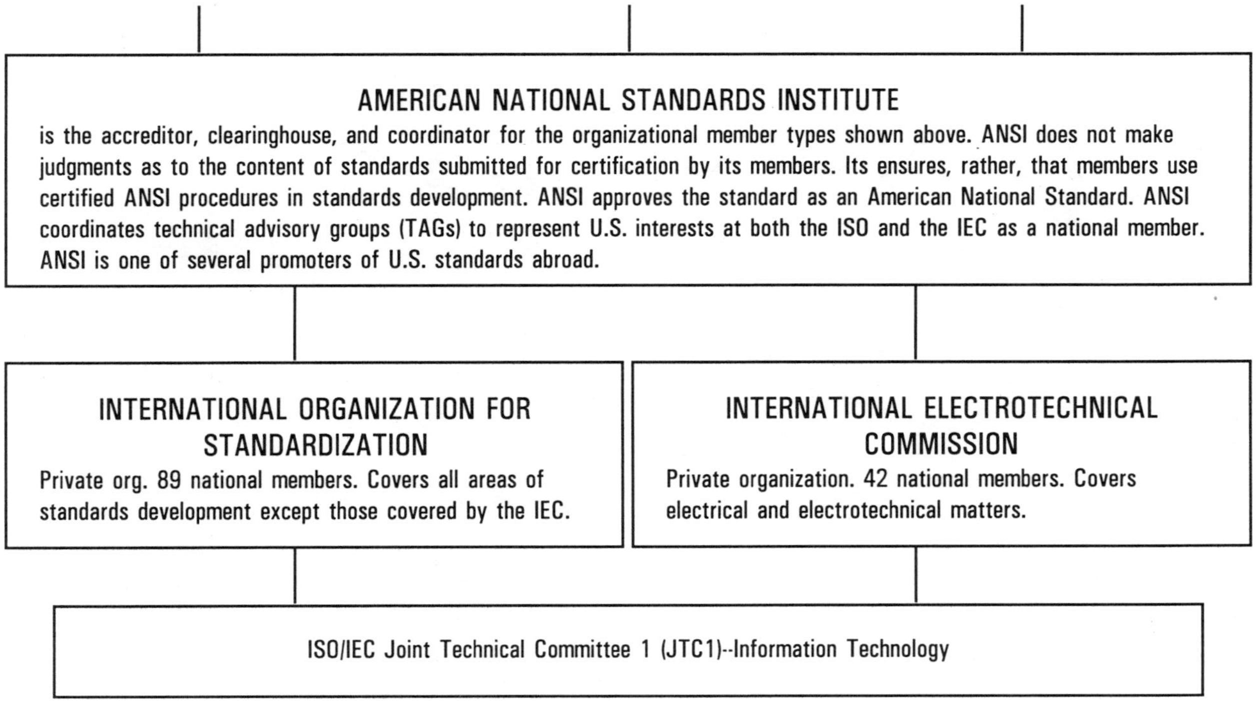

AMERICAN NATIONAL STANDARDS INSTITUTE

is the accreditor, clearinghouse, and coordinator for the organizational member types shown above. ANSI does not make judgments as to the content of standards submitted for certification by its members. Its ensures, rather, that members use certified ANSI procedures in standards development. ANSI approves the standard as an American National Standard. ANSI coordinates technical advisory groups (TAGs) to represent U.S. interests at both the ISO and the IEC as a national member. ANSI is one of several promoters of U.S. standards abroad.

INTERNATIONAL ORGANIZATION FOR STANDARDIZATION

Private org. 89 national members. Covers all areas of standards development except those covered by the IEC.

INTERNATIONAL ELECTROTECHNICAL COMMISSION

Private organization. 42 national members. Covers electrical and electrotechnical matters.

ISO/IEC Joint Technical Committee 1 (JTC1)--Information Technology

FIGURE 2-3 ANSI-accredited standards development process.

obtain ANSI recognition for a standard by submitting it to an open ballot—a canvass—to verify agreement among parties who would be directly affected by the proposed standard. ANSI accreditation and publication of these standards as American National Standards enables them to achieve similar status and distribution—and thus compete for influence and sales—with those of SDOs that use a more thorough, committee-based consensus process. ANSI currently accredits 112 canvass sponsors, 42 organizations, and 199 standards committees.[59]

It is difficult to quantify either the extent of competition or the relative merits of canvass and committee standards. Such a determination is beyond the direct scope of this report. The circumstances of each case vary widely by industry, product, and type of standard. In the committee's judgment, nevertheless, this conflict is not a threat to the viability of the U.S. private standards development system. Whereas the leading SDOs develop and maintain hundreds of standards, most canvass developers are responsible for fewer than five standards.[60] In addition, recent progress in resolving this and other conflicts between ANSI and leading SDOs appears to provide a basis for optimism that the private-sector, voluntary standardization system is capable of settling internal disagreements and continuing to meet the nation's need for standards.

A second challenge to ANSI, and to the U.S. consensus standardization system as a whole, has been the rapid advance of technology in some industry sectors. Slowness of consensus standards processes is a widely cited problem.[61] Agreement among competing firms on the best technical standard for a given purpose can be difficult to achieve. Basic communication about technical questions may be costly and time-consuming, requiring numerous technical committee meetings and frequent correspondence. Due process requirements may add delay. Legitimate differences of technical opinion may be compounded, moreover, by participants' competition for marketing advantage. For example, a firm may have an overall interest in standardization, but seek delay in an effort to ensure that no standard is adopted, rather than allowing a competitor's technology to become the standard.[62]

Technological uncertainty compounds the difficulty of writing standards. To keep up with technological change, technical committees increasingly must set *anticipatory standards*. These are standards that describe technologies and products not yet completely developed. Recent economics research, moreover, suggests that setting compatibility standards for rapidly evolving information and telecommunications technologies presents a unique challenge to consensus standards developers. The challenge of compatibility standards arises from two phenomena. The first is the potential for significant economic rewards to firms that succeed in setting a proprietary, de facto compatibility standard, such as a computer operating system. The second is the possibility of a technology bandwagon, in which users rush to adopt a standard once it appears that most other users will adopt that standard. The corollary of the bandwagon is technology lock-in. Once most users have committed to a compatibility standard, there are significant costs to switching to a new standard—even if it represents a more advanced or useful technology.[63]

Bandwagon and lock-in effects can reward some technologies with large market shares and hinder the success of others, irrespective of their technical merits. Lock-in of inferior technologies through standards can, in some instances, retard innovation and technological advance. These phenomena can occur faster than the typical development cycle of consensus standardization. The rise of standards consortia in the information technology and telecommunications industries is one response to this challenge. The limited due process, consensus, and open participation requirements of these organizations enable them to develop standards rapidly in many instances. Their procedures and restricted membership, however, may limit the acceptability of these standards outside the consortia that develop them.[64] Consensus standards developers are responding to this challenge with such measures as streamlined due process and a tighter focus on customer needs in setting the scope for standards writing. In the past five years, for example, the international consensus standards developer for information technology, ISO/IEC Joint Technical Committee 1 (JTC1), has reduced the time needed to produce an international standard from more than 50 months to less than 36 months.[65]

The best means to achieve standardization, in the committee's judgment, is a flexible, sector-specific approach. Issues such as appropriate speed, technological sophistication, openness of participation, and degree of consensus for standards should be determined by participants in each industry sector. Standards development cycles that are too slow for the telecommunications industry, for example, might be too fast for building materials or consumer appliances. No single set of SDO procedures or performance criteria is likely to meet the needs of manufacturers and users across technologically and economically diverse industry sectors.

Common to all industry sectors, however, is the need for greater accessibility of information about standards and standardization processes. As noted previously, coordination costs are a significant hurdle for achieving standardization. Numerous SDOs with formal procedures for convening technical experts have come into existence in order to overcome this hurdle. Modern communications technologies, however, present additional opportunities to reduce the barriers to participation in standards development, particularly for small firms with limited resources. The National Standards Systems Network (NSSN), a pilot program administered by ANSI under a $2 million cooperative agreement with NIST, is intended to foster links among existing sources of standards information. Electronic dissemination is a key element of NSSN.[66] Additional efforts of this type hold significant potential for facilitating participation, particularly for small enterprises and consumer interests. Other benefits will include lowering costs and increasing the speed and efficiency of the U.S. standards development system.

A third area in which ANSI's role has evolved through periods of both tension and cooperation is its relationship with standards developers and users in the U.S. government.[67] As a matter of policy, federal agencies are committed to adopt voluntary consensus standards to the greatest possible extent, rather than developing new, government-unique standards. In the final section of the chapter, cooperation and sources of tension between ANSI and the U.S. government related to federal participation in the voluntary standards system are discussed. In the next section, however, U.S. participation in international standards development through ANSI and other avenues is examined.

INTERNATIONAL STANDARDS DEVELOPMENT

The two predominant international standards-setting bodies in the world are the International Organization for Standardization and the International Electrotechnical Commission. ISO and IEC are private organizations that develop standards in nearly all sectors of industry and technology. The largest exception to their coverage is international telecommunication standardization, which is the domain of the International Telecommunications Union (ITU). ITU is a treaty organization with membership comprised of government representatives from 160 countries. U.S. representation at ITU is coordinated by the Department of State.[68] As private agencies, ISO and IEC accept as members the national standards organizations, whether public or private, of their member countries. ANSI is the U.S. national member of ISO and IEC, the latter through the ANSI-coordinated U.S. National Committee. (See Box 2-1 for additional background information and comparisons of ISO, IEC, and ITU).

International standards development processes resemble those of U.S. private SDOs in many respects.[69] ISO and IEC prepare standards within a decentralized technical committee structure, drawing on volunteer technical experts from various member countries. Administrative support for technical committees is provided by a secretariat, from one of the participating countries. Standards are drafted through consensus. Voting within committees and in the organization at large, unlike many national SDOs, is by national delegation. As a result, a large country such as the United States has the same vote as a small country.

U.S. positions for international standardization activities are developed by volunteer experts within technical advisory groups (TAGs). ANSI coordinates the formation of U.S. TAGs corresponding to technical committees at the international level. In addition, on issues that the U.S. standards community considers of particular importance—such as standards affecting large shares of U.S. exports—ANSI and the U.S. standards community make efforts to obtain ISO and IEC designation of the United States as the secretariat for particular international technical committees.

The United States is a participant or observer in 95 percent of ISO, and nearly all IEC, technical committees and subcommittees.[70] The United States held 13.7 percent of ISO and IEC technical committee and subcommittee secretariats in 1992, an increase from 10.9 percent in 1988.[71] These committees and subcommittees are concentrated in especially active areas of standards activity, producing in 1991 more than 38 percent of all ISO and IEC standards and 31 percent (measured in pages of text) of Draft International Standards. (The latter represents a significant increase over a 6.8 percent share in 1988.[72])

In addition, the United States has had significant success in obtaining secretariats of ISO and IEC technical committees and subcommittees in industry sectors with high volumes of exports. For example, the United States holds the secretariats of ISO/IEC JTC1 for Information Technology; ISO Technical Committee (TC) 20, covering aircraft and space vehicles; ISO TC 61, plastics; and ISO TC 67, petroleum industry materials and equipment, among others. All of these committees set international standards in industry sectors that are among the top 10 U.S. export industry sectors.[73]

Cooperation between the U.S. public and private sectors—which is discussed in depth in the next section of this chapter—was instrumental in gaining a strong U.S. role in the recent establishment of an ISO technical committee on sterilization of health care products, an area of interest to U.S. exporters in the medical devices industry. Coordination among the AAMI, the FDA, and the Health Industry Manufacturers Association (HIMA) was instrumental in ANSI's developing a successful proposal and gaining ISO approval for a new committee, ISO TC 198. The international secretariat was assigned to the United States, where it is sponsored and staffed by the AAMI. A key goal of AAMI in pursuing this outcome, according to its staff, was to be able to cooperate with European standards developers to ensure harmonization of U.S., European, and international regulations.[74]

The scope of international standardization is broader than ISO and IEC alone. U.S. participation in ITU is significant to international standards development in the telecommunications equipment and services industries. In addition, standards produced by some U.S. standards developers take on the authority of international standards without going through a process of consensus building at ISO and IEC. For example, ASTM standards are used throughout the world, and 33 percent of ASTM's sales of publications are outside the United States.[75] About 20 percent of ASME's sales of codes and standards are non-U.S. sales.[76]

The significance of both international standards and conformity assessment is growing in conjunction with the increasing importance of international trade to U.S. economic performance. Expansion and strengthening of international trading system rules concerning standards and conformity assessment provide additional incentives to U.S. industry, government, technical, and other participants in the standards system to focus increased attention on international activities. These and related factors are examined in detail in Chapter 4.

BOX 2-1
INTERNATIONAL STANDARDS DEVELOPERS

International Organization for Standardization

The ISO is a private international agency, established in 1946 with headquarters in Geneva, Switzerland. It is dedicated to voluntary standardization. Its membership consists of recognized national standards organizations of 89 countries. ISO covers work in all areas of standards development except those in the fields of electrical and electrotechnical standards, the domain of the IEC, and telecommunications, the expertise of the ITU. ISO has more than 160 technical committees, some 600 subcommittees, and a host of working groups that are supported by secretariats in 32 countries. The development process is lengthy and ultimately requires the majority consensus of technical committee members and 75 percent of the ISO voting membership. Only after consensus has been reached is it published by the ISO Council as an International Standard. The American National Standards Institute is the U.S. representative to ISO. ISO has published more than 6,700 international standards since its inception.

International Electrotechnical Commission

The IEC, an international voluntary organization headquartered in Geneva, Switzerland, specializes in standards development for electrical and electronic engineering. IEC is concerned mostly with creating specification standards for products and devices. It has a membership of presidents from the national committees of 42 countries. The development process is a lengthy one. The technical work is done by about 200 technical committees, which are managed by the Committee for Action. This committee has three advisory committees; an Advisory Committee on Electronics and Telecommunications, an Advisory Committee on Safety, and the Information Technology Coordinating Group. IEC issues publications and recommendations for international standards, as well as promoting safety, compatibility, interchangeability, and acceptability. To keep nonmember countries informed of the process and development of its standards, IEC created the Registered Subscriber Service. ANSI is the U.S. representative to the IEC. A major difference between ISO and IEC is that in the IEC each member nation has membership on every technical committee not just on those it chooses to join.

GOVERNMENT ROLE IN STANDARDIZATION

The public sector plays a major part in the U.S. standards system. Federal, state, and local government agencies are all active in developing and using standards. Standards written by federal agencies for regulatory and procurement purposes comprise more than half of the total number of U.S. national standards, as shown in Table 2-1. These are categorized as mandatory standards, reflecting their imposition through legislation and regulation or through contractual requirements for sale to government purchasers. Although these standards are developed outside the ANSI-coordinated voluntary consensus system, the mandatory and voluntary standards categories overlap. Many government standards

International Telecommunication Union
The ITU is the only international standards development organization that is non-voluntary. It is a treaty organization run under the auspices of the United Nations. Governments, not industry, administer and enforce the regulatory telecommunications standards that come out of the ITU. It has a membership of 160 nations. The U.S. representative to the ITU is the State Department. A Plenipotentiary Conference acts as the authority and sets policy, while the council, composed of 43 members elected by the conference, deals with administrative matters. ITU maintains five permanent activities. They are the General Secretariat, the organization of World Conferences on International Telecommunications, the Radiocommunication Sector, the Telecommunications Standardization Sector, and the Telecommunications Development Sector. The ITU typically develops recommendations that are implemented as national standards by national telecommunications authorities.

ISO, IEC, and the ITU are alike in two important ways. They all have similar administrative structures with committees, subcommittees, and working groups directing the standards-setting process, with varying levels of complexity. Also, they all rely on some form of consensus as the ultimate decision-making mechanism. None of the three organizations demands or expects participation from all nations; participation is voluntary at all levels. However, once a nation becomes a member it is expected to be actively involved. The focus on consensus is aimed at preventing a decentralization at the international level of standards development and encouraging broad-based compliance and harmonization.

SOURCES:
"International Standards: It's a Small World After All." In *Quality*, Wheaton, IL: Hitchcock Publishing Co., August 1986.

Cargill, Carl F. *Information Technology Standardization: Theory, Process, and Organizations.* Pps. 126-145. Bedford, MA: Digital Press, 1989.

International Telecommunications Union informational brochure. Geneva: ITU Public Relations, 1993.

make reference to consensus standards in whole or in part. This process has the effect of making many voluntary consensus standards, in effect, mandatory.

The Department of Defense and the General Services Administration (GSA), with 38,000 and 6,000 procurement standards, respectively, represent the bulk of federal standards development. The remaining 8,500 standards, mainly technical regulations, are produced by a wide range of departments and agencies (see Table 2-3). Regulatory standards center on the protection of public health and safety. Examples include regulations set by the Food and Drug Administration, Occupational Safety and Health Administration (OSHA), Consumer Product Safety Commission (CPSC), and Federal Aviation Administration, among other agencies. The Environmental Protection Agency (EPA) regulates products and processes

TABLE 2-3 — U.S. Government Standards Developers

Agriculture, Department of
Agricultural Marketing Service
Federal Grain Inspection Service
  Field Management Division
    Standards and Procedures Branch
Food Safety and Inspection Service
Foreign Agricultural Service
Forest Service
  Engineering Staff
Information Resources Management
  Planning, Review, and Standards Division
Packers and Stockyards Administration
  Livestock Marketing Division
Rural Electrification Administration

Commerce, Department of
Bureau of the Census
Federal Coordinator for Meteorology
National Institute of Standards and Technology
  National Computer Systems Laboratory
  National Engineering Laboratory and Law
  Enforcement Standards Laboratory
  Technology Services - Voluntary Product
  Standards
National Oceanic and Atmospheric
Administration
  National Marine Fisheries Service
  National Environmental Satellite, Data, and
  Information Service
  National Weather Service
National Telecommunications and Information
Administration
  Institute for Telecommunications Sciences
U.S. Patent and Trademark Office
  Assistant Commissioner for Information
  Systems
  Assistant Commissioner for Patents
  International Patent Documentation
  Trademark Examining Operation

Consumer Product Safety Commission
Directorate for Engineering Sciences
Directorate for Health Sciences

Defense, Department of
Office of the Assistant Secretary of Defense,
Acquisition
Defense Industrial Supply Center

Energy, Department of
Assistant Secretary for Defense Programs
Building Technologies
  Building Systems and Materials Division
  Building Equipment Division

Energy Information Administration
  Statistical Standards
Environment, Safety, and Health
  Safety and Quality Assurance

Environmental Protection Agency

Federal Communications Commission
Office of Engineering and Technology

General Services Administration
Information Resources Management
Federal Supply Service
  Commodity Management
Public Building Service

Health and Human Services, Department of
Centers for Disease Control
  National Institute for Occupational Safety
  and Health
Food and Drug Administration
  Regulatory Affairs
Health Care Financing Administration

**Housing and Urban Development, Department
of**
Assistant Secretary for Housing - Federal
Housing Commissioner
  Manufactured Housing and Construction
  Standards Division

Interior, Department of the
Minerals Management Service
  Rules, Orders, and Standards
U.S. Geological Survey
  Information Systems Division
  National Mapping Division
  Water Resources Division

Labor, Department of
Mine Safety and Health Administration
  Standards, Regulations and Variances
Occupational Safety and Health Administration
  Directorate of Safety Standards Programs

National Aeronautics and Space Administration
Occupational Health
Safety, Reliability, Maintainability, and Quality
Assurance Division

National Archives and Records Administration
Archival Research and Evaluation Staff

Nuclear Regulatory Commission
Nuclear Regulatory Research

TABLE 2-3 — (continued)

State, Department of
U.S. National Committee for the International Telecommunications Union-Telecommunication Standardization Sector

Transportation, Department of
Federal Aviation Administration
Federal Highway Administration
Maritime Administration
National Highway and Traffic Safety Administration
Research and Special Programs Administration
  Standards Division
United States Coast Guard
  Marine Safety, Security, and Environmental Protection
  Auxiliary, Boating, and Consumer Affairs Division

Treasury, Department of
Bureau of Alcohol, Tobacco, and Firearms
  National Laboratory Center
Internal Revenue Service
  Standards and Data Administration
U.S. Customs Service
  Commercial Operations
  Research Division - Laboratories and Scientific Services

Veterans Affairs, Department of
Acquisition and Material Management

SOURCE: Toth, Robert B. *Standards Activities of Organizations in the United States*. NIST Special Publication 806. U.S. Department of Commerce. Gaithersburg, Md.: NIST, 1991.

that affect the environment. The Federal Communications Commission (FCC) sets telecommunications equipment standards to ensure compatibility and to protect the security and integrity of the public communications network. The Department of Agriculture produces standards both to promote food safety and to ensure accurate grading and marketing of agricultural products. The Department of Commerce's National Institute of Standards and Technology, among other standards-related activities, develops and maintains standards for physical measurement, known as reference standards.[77]

Standards-writing activities of state and local governments are less easily identified than those of the federal government. These levels of government are very active in the areas of product certification and laboratory accreditation. Such programs, however, largely make assessments against standards originally written by other authorities—for example, private building codes organizations for construction standards and the NFPA's ANSI-approved National Electrical Code.[78] (The automobile emissions standards written by the State of California are a well-known exception.) The impact of state and local standards-related activities, as discussed in Chapter 3, is reflected primarily in conformity assessment rather than in standards development. In a recent pilot project performed by NIST's National Center for Standards and Certification Information, for example, the official gazettes of California, Texas, and New Mexico were monitored for announcements of new standards development activity that might affect trade within the North American Free Trade Agreement region. Only one standard was identified during several months of 1994.[79]

National Institute of Standards and Technology

The U.S. government agency with leading expertise in the area of technology standards and industry standardization issues is the Department of Commerce's National Institute of Standards and Technology (NIST). Although NIST is not a regulatory or a procurement agency, it is active in many aspects of both public and private standards setting. NIST was established in 1901 as the National Bureau of Standards, with responsibility for developing and coordinating reference standards—standards of weights and measures. In 1988, the bureau was reconstituted as NIST and given the explicit mission of assisting U.S. industry to advance its performance in the development and application of technology. Scientists at NIST's internal laboratories conduct basic and applied research in a wide range of physical sciences. One central goal of this research is to advance the science of measurement and testing and to apply these advances to standardization.[80]

For fiscal year (FY) 1994, NIST's overall budget was $520.2 million.[81] Its FY 1995 appropriation request increased to $935.0 million (see Table 2-4). Most of the increase was in NIST's extramural industry assistance programs, including the Advanced Technology Program, the Manufacturing Extension Partnership, and the Quality Program. Appropriations for intramural programs, consisting primarily of the eight NIST laboratories, grew from $226.2 million to $316.0 million. The Office of Standards Services—which leads NIST's interactions with the voluntary consensus standards community and other federal agencies with standards activities—is part of NIST's Technology Assistance activity. The appropriation for Technology Assistance, of which the Office of Standards Services accounts for about half,[82] grew from $11.0 million in FY 1994 to $14.9 million in FY 1995. The Office of Standards Services therefore represents approximately one-half of 1 percent of NIST's overall budget.

NIST is not a regulatory or a procurement agency, and it does not set mandatory standards. (An exception is the Federal Information Processing System [FIPS] standards series, procurement specifications for federal data systems developed under NIST leadership.[83]) NIST staff are, however, highly involved in both U.S. and international voluntary consensus standards development. In 1993, 380 members of NIST's research laboratory staff participated in consensus standards committees. The committees were associated with 59 domestic and 20 international SDOs. Almost half of these committees were associated with ASTM, reflecting NIST's particular expertise in testing and measurement. Participating staff members held an average of three committee memberships.[84] The number of full-time equivalent staff represented by these activities was not reported. In 1991, however, NIST staff reported 31,787 labor hours for travel and

TABLE 2-4 — NIST Appropriation Budget Summary by Subactivity—FY 1993 - 1995 (millions of dollars)

	FY 1993 Approp.	FY 1994 Approp.	FY 1995 Approp.	FY 1995 (%)
Industrial Technology Services (Extramural Programs)				
Advanced Technology Program	67.9	199.5	430.7	50.4
Manufacturing Extension Partnership	18.2	30.2	90.6	10.6
Quality Program	-----	2.8	3.4	0.4
Scientific and Technical Research and Services (Intramural Programs)				
Technology Assistance (including Standards Services)	**8.5**	**11.0**	**14.9**	**1.7**
Electronics and Electrical Engineering	26.5	29.5	35.4	4.1
Manufacturing Engineering	10.1	13.6	19.2	2.2
Chemical Science and Technology	19.3	22.2	32.5	3.8
Physics	26.4	26.7	27.5	3.2
Materials Science and Engineering	35.6	43.3	49.8	5.8
Building and Fire Research	12.0	12.8	13.2	1.5
Computer Systems	12.1	28.9	37.1	4.3
Applied Mathematics and Scientific Computing	6.8	7.0	7.2	0.8
Research Support Activities	35.6 [a]	31.2 [a]	27.5	3.2
Construction of Research Facilities				
Construction and Major Renovations	105.0	61.7	64.6	7.6
Total NIST Appropriations	384.0	520.2	853.8	100.00

[a] Includes portions of facilities funding that are included in the Construction of Research Facilities appropriation in FY 1995.

SOURCE: Budget Office, National Institute of Standards and Technology, U.S. Department of Commerce, 1994.

participation in domestic and international standards committees, at a total cost of more than $1 million.[85]

Compilation of data on NIST staff's participation in consensus standards setting is one among several functions of the Office of Standards Services. The Director of Standards Services chairs the federal Interagency Committee on Standards Policy, discussed below. The office also serves as the U.S. inquiry point for standards within the international trading system, the General Agreement on Tariffs and Trade (GATT). Through a network of national inquiry points, GATT members are required to notify each other when considering new regulations and conformity assessment requirements that affect imports from other nations. (GATT obligations concerning standards and conformity assessment are discussed in detail in Chapter 4.) The office is also the U.S. contact point for ISONET, an information exchange network for members of the International Organization for Standardization, despite the fact that ANSI is the U.S. member body of ISO. Other Office of Standards Services activities discussed in the next two chapters include laboratory accreditation and conformity assessment system recognition; coordination of the National Conference on Weights and Measures, which promotes uniformity and effectiveness in state and local measurement programs; technical assistance programs in several developing overseas markets; and assistance to the U.S. Trade Representative and other trade agencies with technical standards and conformity issues that affect international trade policy.[86]

In conjunction with its role as the U.S. GATT and ISONET inquiry points, the Office of Standards Services maintains an extensive library of information about both U.S. and international standards, including mandatory and voluntary standards. This library, the National Center for Standards and Certification Information (NCSCI), is open to the public, responds to telephone and written inquiries, and disseminates standards information through announcements in the *ANSI Reporter*. Through its Standards Code and Information (SCI) Program, Standards Services compiles directories of public and private organizations with standards and conformity assessment activities and publishes basic informational reports on various topics. With a staff of 10 and a combined annual budget of approximately $1 million, however, SCI and NCSCI have been unable to pursue as proactive an outreach effort as would be possible with greater resources. For example, NCSCI receives approximately 10,000 requests for information per year. These divide about evenly between questions about domestic and foreign standards and conformity assessment matters. Increased efforts at publicizing the service—for example, through advertisements in industry and trade publications—would likely swamp the center's capacity to respond to inquiries.[87] A 1993 special publication from SCI, a report on ISO 9000 quality system standards, became its most requested document; however, lack of resources for printing and mailing has limited dissemination of the report.[88]

Federal Use of Voluntary Consensus Standards

Many federal agencies besides NIST are active in developing and using standards. Procurement standards set by the DoD and the GSA together represent the majority of federal standards. Regulatory agencies such as EPA, OSHA, CPSC, and FDA account for approximately 8,500 active federal standards. Federal regulations and procurement standards are distinct from voluntary consensus standards in two respects. First, they are mandatory—by law, by regulation, or by contractual obligation in government purchasing. Second, federal standards are not generally written by committees of volunteer experts through a consensus-seeking procedure. Administrative procedures law requires public notice of proposed rules in the *Federal Register* and response to received comments, but agencies' statutory obligations require government employees to make any final decisions in setting government standards.

Increasingly, however, government agencies are meeting their statutory obligations not by developing government-unique standards but, rather, by participating in and adopting the end products of voluntary consensus standards development. In 1982, the Office of Management and Budget (OMB) issued Circular A-119, "Federal Participation in the Development and Use of Voluntary Standards."[89] Revised in October, 1993, Circular A-119 notes,

> Government functions often involve products or services that must meet reliable standards. Many such standards, appropriate or adaptable for the Government's purposes, are available from private voluntary standards bodies. Government participation in the standards-related activities of these voluntary bodies provides incentives and opportunities to establish standards that serve national needs, and the adoption of voluntary standards, whenever practicable and appropriate, eliminates the costs to the Government of developing its standards. Adoption of voluntary standards also furthers the policy of reliance upon the private sector to supply Government needs for goods and services, as enunciated in OMB Circular No. A-76, entitled *Performance of Commercial Activities.*[90]

The policy expressed in Circular A-119 has strong potential to produce savings to government in developing standards. Participation of government experts such as NIST research staff in consensus standards committees raises the level of technical competence applied to the standardization effort. Committees serve as a working forum for public-private cooperation in the development of standards to meet public needs, while imposing the lowest possible costs and restrictions on technological innovation in industry. The circular encourages government use of performance standards over design standards for this reason.[91]

In public procurement, use of consensus standards in place of government-unique specifications has proven effective at both reducing government costs and improving the competitive strength of U.S. industry. Pilot efforts at DoD in replacing military with commercial item specifications have saved procurement funds and reduced burdens on suppliers of maintaining separate commercial and military production capabilities.[92] One example is the procurement of thermal insulation for buildings. The Naval Facilities Engineering Command (NAVFAC) reviewed government and consensus standards in this area and found many redundant standards.[93] In 1982, at NAVFAC's request, ASTM formed a technical committee to help convert military and civilian federal standards for thermal insulation to ASTM standards. Of an identified 59 candidate government specifications, 33 had been canceled by June 1991. Among these, 20 were replaced by ASTM standards, 12 were canceled without replacement, and one military specification was canceled as duplicative of a civilian federal specification.

Total administrative savings to the Navy of these canceled government-unique specifications were estimated at more than $1.8 million. NAVFAC also expects to save about 2 percent on material cost of insulation. Now that appropriate consensus standards have been identified, these savings should be replicable across all federal military and civilian purchases of insulation. This will yield estimated total savings to the U.S. government of $89.5 million over the life-cycle of a typical facility. In addition, the defense supplier base for insulation has been strengthened by the conversion to commercial items, because administrative duplication and special knowledge related to military purchasing procedures are no longer required of suppliers.[94]

Controlling costs is clearly not the only issue faced by federal agencies in setting standards. Regulators, in particular, face an obligation to protect the public interest that sometimes outweighs the need to promote either government or industry efficiency. Circular A-119, moreover, requires use of voluntary standards only "whenever practicable and appropriate." These circumstances are not, however, rare. For several reasons, adoption of voluntary consensus standards by federal agencies—particularly, but not exclusively, when government personnel have participated in their development—is an effective means of securing public interests. First, although voluntary standards-setting is sometimes criticized for slowness, regulatory standards-setting is even slower.[95] Agencies face stringent due process requirements and opportunities for private interests to delay regulatory action through the legal system, as well as limitations on time and resources for drafting regulations. With the exception of especially hazardous product sectors such as drugs, moreover, agencies have generally been far more effective at influencing corporate design and production of safe products through public information campaigns, including product advisories, and product recalls than through the writing of mandatory standards.[96]

Second, voluntary consensus standards are often equally as stringent in the level of protection they require as mandatory standards would be.[97] It might seem reasonable to expect that private standards developers—industry associations, especially—would seek to set standards at the lowest common denominator of safety. Such standards might allow manufacturers to cut costs, for example. In fact, however, private standards writers have several incentives to set high standards. Forestalling government regulation by developing a private solution to a perceived problem requires a standard stringent enough to satisfy public needs. (Government participation in standards committees enhances this process from both public and private perspectives.) Voluntary design of safe products also reduces risks of large liability claims and high liability insurance premiums. Avoidance of liability has been found, in fact, to be a stronger motivator for safe product design than regulation or any other factor.[98] The private standards system plays an important part in guiding corporate decision making and disseminating safe and in effective design throughout industry.

The potential benefits of effective public–private cooperation in the development and use of consensus standards are significant. Implementation of OMB Circular A-119 has not, unfortunately, been successful at securing these benefits. The NIST-chaired Interagency Committee on Standards Policy (ICSP) has a mandate to coordinate policy throughout the federal government on using voluntary consensus standards. In 1987, because of a lack of commitment on the part of regulatory agencies, ICSP was virtually disbanded.[99] The OMB's 1993 revision of the circular required each agency to appoint an senior Standard Executive to serve on the ICSP, and it raised the frequency of required reports on agency activities in voluntary standardization from triennial to annual.[100] The ICSP was rechartered in June 1991 and has begun meeting approximately annually. Table 2-5 lists the 1994 membership of the ICSP, including title and department. Working groups under ICSP have been active in the following areas: compiling directories of agency staff participation in standards committees; federal use of ISO 9000 quality standards; conformity assessment (from February to December 1992); and international standards (from March to September 1992).[101]

The 1992 triennial report of the ICSP, however—prepared after revision of Circular A-119 had been initiated—noted the following:

"Despite the low level of committee activity, significant standards-related activities are underway in a number of agencies, albeit in an uncoordinated fashion. . . . Having only one ICSP representative from a department with multiple agencies and the diversity of programs in certain agencies make oversight and accountability difficult."[102]

The report illustrates a high degree of variability among agencies in implementing OMB policy. While DoD reported an increase in the number of voluntary standards used from 3,486 in 1985 to 5,200 in 1991, CPSC use of such standards rose from 6 to 9. In the same period, FCC reported a decline from 6 to 5 voluntary standards used.[103] For a number of agencies, such as the Department of Transportation and the Department of Agriculture (1991), no data were reported.

Public–Private Cooperation

Effective public–private cooperation in developing and using consensus standards will require increased commitment within agencies and improved sharing of information among agencies and between the public and private sectors. The revised reporting and membership requirements in Circular A-119 are unlikely to achieve the needed improvements. The circular makes NIST the chair of the interagency coordination process, but it does not give NIST or any agency a clear mandate to oversee and evaluate federal implementation of the policy across all agencies. The OMB retains final authority for overseeing its policies, but lacks

TABLE 2-5 — Interagency Committee on Standards Policy
as of June 29, 1994

AGENCY MEMBER	REPRESENTATIVE
Chair - Commerce, Department of - National Institute of Standards and Technology	Director, Office of Standards Services, NIST
Agency for International Development, U.S.	Director, Office of Administrative Services
Agriculture, Department of	Director, Office of Food Safety and Technical Services
Consumer Affairs, Office of	Director for Policy and Education Development
Consumer Product Safety Commission	Assistant Executive Director for Hazard Identification and Reduction
Defense, Department of	Deputy Assistant Secretary (Production Resources)
Education, Department of	Chief Financial Officer
Energy, Department of	Director, Office of Nuclear Safety Policy and Standards
Environmental Protection Agency	Deputy Director, Office of Modeling, Monitoring Systems and Quality Assurance
Federal Communications Commission	Chief Engineer
Federal Emergency Management Agency	Deputy Associate Director, Operations Support Directorate
Federal Trade Commission	Associate Director for the Bureau of Consumer Protection
General Services Administration	Deputy Assistant Commissioner, Office of Commodity Management, Federal Supply Service
Government Printing Office, U.S.	Manager, Quality Control and Technical Department
Health and Human Services, Department of - Food and Drug Administration	Director, Office of Standards and Regulations, Center for Devices and Radiological Health
Housing and Urban Development, Department of	Senior Advisor for Science, Technology and Urban Policy
Interior, Department of the	Director, Office of Acquisition and Property Management
International Trade Commission	Director, Office of Administration
Justice, Department of	Director, Office of Policy Development
Labor, Department of	Assistant Secretary for Administration and Management
National Aeronautics and Space Administration	Associate Administrator for Safety and Mission Assurance
National Archives and Records Administration	Preservation Officer

TABLE 2-5 — (continued)

AGENCY MEMBER	REPRESENTATIVE
National Communications System	Assistant Manager for Technology and Standards
National Science Foundation	Senior Engineering Advisor
Nuclear Regulatory Commission	Deputy Director, Division of Engineering, Office of Nuclear Regulatory Research
Postal Service, U.S.	Manager, Configuration Management
Small Business Administration	Deputy to the Associate Deputy Administrator for Economic Development
State, Department of	Deputy Assistant Secretary for Trade and Commercial Affairs
Trade Representative, U.S.	Deputy Assistant U.S. Trade Representative for GATT Affairs
Transportation, Department of	Director, Office of International Transportation and Trade
Treasury, Department of	Deputy Assistant Secretary for Information Systems
Veterans Affairs, Department of	Deputy Assistant Secretary for Acquisition and Material Management
Liaison - Office of Management and Budget	Chief, Information Policy Branch, Office of Information and Regulatory Affairs
Executive Secretary - National Institute of Standards and Technology	Standards Code and Information Program, Office of Standards Services

the depth of expertise in standardization issues available to NIST. An oversight mandate for NIST would lead to improved federal use of voluntary standards and enhancement of the government's regulatory and procurement activities at reduced cost.

A NIST mandate would also create a central locus for coordinating communications on standards issues between the federal government and the private standards community. Despite many discussions over past decades, there has never been a formal Memorandum of Understanding (MOU) between ANSI and the U.S. government.[104] An MOU would lay out the respective roles and responsibilities of government and the private sector in the U.S. standards system. It would not make ANSI the officially designated developer of standards for the U.S. government; final authority for protecting public interests must, by law, remain with federal agencies. An MOU would, however, facilitate understanding throughout the government of the potential uses of voluntary standardization in meeting public objectives. It would recognize the system of voluntary consensus standardization, conducted by SDOs with coordination and accreditation by ANSI, as a valuable source of standards for public use.

An MOU would affirm ANSI's responsibility and improve its ability to represent U.S. interests in international, nontreaty standards-setting bodies. Although the Trade Agreements Act of 1979 recognizes that U.S. representation in international standardization should be by the private U.S. member of the relevant organization, it does not specify mechanisms for government cooperation with ANSI and U.S. industry in preparing U.S. positions for international standards activities.[105]

An MOU would also be an appropriate vehicle for addressing a frequent source of tension in public–private standards cooperation. This source is the low level of government financial support for voluntary standards organizations, including ANSI. Government agencies make significant contributions to voluntary standardization, as shown by the previously discussed data on NIST staff participation in outside standards committees. ANSI incurs significant expenses, however, in providing the administrative overhead for coordinating the U.S. voluntary standards system. ANSI dues to ISO and IEC are a particularly large expense.

Government participation and use of the system implies a responsibility to pay a share of the overhead expenses associated with the system. Among federal agencies, however, only the Departments of Agriculture, Defense, Energy, and Veterans Affairs, along with NIST, the U.S. Geological Survey, FCC, FDA, GSA, the NASA, and the National Archives are dues-paying members of ANSI. EPA, CPSC, the Nuclear Regulatory Commission, and the Departments of Housing and Urban Development, Labor, and Transportation are among prominent government standards developers that are not ANSI members.[106] Although membership dues represent 23 percent of ANSI gross revenues (including publications sales), U.S. government dues are less than half of 1 percent of ANSI revenues.[107]

Cooperation and understanding between the private standards system and the federal government appear to be improving. A formal MOU clearly would not create understanding where none exists. It would, however, create a formal framework for continuation of cooperation in the future. This framework would prove valuable as circumstances change, new issues emerge, and informal working relationships among individuals in each sector are replaced through changeover of key personnel.

SUMMARY AND CONCLUSIONS

The U.S. system for developing formal product and process standards is complex and diverse. It incorporates, for example, cooperative efforts by technical experts to write voluntary standards on a consensus basis. These activities generally take place in the context of consensus-based standards-developing organizations, according to guidelines for due process and open participation of interested parties. Many standards developers are accredited by the American National Standards Institute, a private, nonprofit federation of business, government, and other individuals. Other U.S. standards arise through competition in the marketplace. When a particular set of product specifications is widely used, it may become a de facto market standard. Government agencies set mandatory standards, both to meet regulatory needs in areas such as health and safety and to support public-sector procurement of products and services. Voluntary standards developed in the private sector may become mandatory through adoption as government standards.

This chapter has presented analysis and evidence to demonstrate that the U.S. standards development system is largely successful in meeting public and private needs for standards. The U.S. standards system has a highly decentralized structure. It offers multiple avenues for developing standards and disseminating them to potential users in industry and government, including informal (de facto), consensus-oriented, and mandatory processes. These characteristics provide for responsiveness to a wide range of demands for particular standards. These demands vary according to such factors as industry structure; level of development and speed of technological change; and specific, relevant public interests, such as protection of health and safety.

As this chapter has noted, however, there is the need for significant improvement in the use of standards developed in the private sector for government regulation and procurement. Government use of private standards has the strong potential to reduce costs for government agencies and private industry. Existing federal policies, however, are ineffective at ensuring that these benefits are realized. New mechanisms are needed to provide for (1) improved standards policy coordination with the federal government and (2) improved communication and cooperation between government and private sector standards organizations, particularly ANSI. Chapter 5 presents specific recommendations for achieving these needed improvements.

Processes for developing standards represent only a portion of the complete impact of standards in the U.S. economy. The next chapter examines public and private mechanisms in the United States for assessing the conformity of products and industrial processes to standards.

NOTES

1. Michael Hergert, *Technical Standards and Competition in the Microcomputer Industry.*
2. Paul A. David, *Clio and the Economics of QWERTY*, 332-337.
3. Breitenberg, *The ABC's of Certification Activities in the United States*; and National Fire Prevention Association (NFPA), *National Electrical Code 1993*. For a detailed analysis of regulatory use of private standards, with several case studies, see Cheit, *Setting Safety Standards.*
4. See William Lehr, *Standardization: Understanding the Process*, 550-555.
5. Toth, *Standards Activities of Organizations in the United States*, 5.
6. Toth, *Standards Activities of Organizations in the United States*, 547-548.
7. Toth, *Standards Activities of Organizations in the United States*, 3.
8. Toth, *Standards Activities of Organizations in the United States*, 3.

9.   American Society of Mechanical Engineers (ASME), *1992 ASME Boiler & Pressure Vessel Code.*

10.   Toth, *Standards Activities of Organizations in the United States*, 3.

11.   Toth, *Standards Activities of Organizations in the United States*, 3.

12.   Breitenberg, ed., *Index of Products Regulated by Each State.*

13.   Breitenberg, *The ABC's of Standards-Related Activities*, 5.

14.   ASME, *1992 Boiler and Pressure Vessel Code*; and OTA, *Global Standards*, 43.

15.   Breitenberg, *The ABC's of Standards-Related Activities*, 3.

16.   For a review of the role of the Society of Automotive Engineers in producing standards for the early automotive industry, see Hemenway, *Industrywide Voluntary Product Standards*, 13-18.

17.   Hemenway, *Industrywide Voluntary Product Standards*, 14-15.

18.   Greenstein, *Invisible Hands and Visible Advisors*, 539.

19.   Greenstein, *Invisible Hands and Visible Advisors*, 539.

20.   Kindleberger, *Standards as Public, Collective and Private Goods*, 384-385.

21.   Kindleberger, *Standards as Public, Collective and Private Goods*, 377.

22.   For a review of the relevant literature, see David and Greenstein, *The Economics of Compatibility Standards: An Introduction to Recent Research*, 3-41.

23.   See, for example, Farrell, *The Economics of Standardization: A Guide for Non-Economists*, 189-198.

24.   OTA, *Global Standards*, 61-69.

25.   Lehr, *Standardization: Understanding the Process*, 550.

26.   OTA, *Global Standards*, 50-51.

27.   Joseph Farrell, *Standardization and Intellectual Property*, 42-43.

28.   See, for example, the American National Standards Institute's (ANSI) guidelines for treatment of patent rights in consensus standardization: ANSI, *Guidelines for Implementation of The ANSI Patent Policy.*

29.   Breitenberg, *Standards Activities of Organizations in the United States*, 2.

30.   Cheit, *Setting Safety Standards*, 23-25.

31.   Lehr, *Standardization: Understanding the Process*, 551.

32.   For further discussion of the dynamics of standards committees, see Cargill, *Information Technology Standardization*, 103-112.

33.   Common characteristics of standards processes among SDOs are considered in detail in Lehr, *Standardization: Understanding the Process.*

34.   Lehr, *Standardization: Understanding the Process.*

35.   Cheit, *Setting Safety Standards*, 15.

36.   Cheit, *Setting Safety Standards*, 187-189.

37.   Cheit, *Setting Safety Standards*, 188. An additional example in which the consensus standard developer is deemed competitive is Eliason Corporation v. NSF. See Pamela Klien, *NSF Wins Court Fight.*

38.   ANSI, *The U.S. Voluntary Standardization System: Meeting the Global Challenge*, 17-22.

39.   Cheit, *Setting Safety Standards*, 23-25; and OTA, *Global Standards*, 49-51.

40.   Defense Systems Management College, *Standards and Trade in the 1990s: A Source Book for Department of Defense Acquisition and Standardization Management and their Industrial Counterparts*, 2.2-2.3; Cheit, *Setting Safety Standards*, 23-25; and OTA, *Global Standards*, 49-51.

41.   Institute of Electrical and Electronics Engineers. *IEEE Standards: Annual Activities Report 1993.*

42.   American Society of Mechanical Engineers. *1992 ASME Bioler and Pressure Vessel Code.*

43.   Cargill, *Information Technology Standardization: Theory, Process, and Organizations*, 170-178.

44. Miller, Michael, J. *Hearing on International Standardization: The Federal Role.*

45. Quality. *International Standards: It's a Small World After All.*

46. American Society for Testing and Materials (ASTM), *ASTM 1992 Annual Report*; and ASTM, *ASTM 1993 Annual Report.*

47. NFPA. *The NFPA Standards-Making System*, n.d.

48. NFPA. *National Fire Protection Association: Fact Sheet.*

49. For a discussion of the rise of standard-setting consortia, see Carl Cargill and Martin Weiss, "Consortia in the Standards Development Process," *Journal of the American Society for Information Science*, 559-565.

50. ANSI, *The U.S. Voluntary Standardization System: Meeting the Global Challenge*, 10-11.

51. *ANSI Annual Report*, 1993, 13.

52. *ANSI Annual Report*, 1993, 1.

53. ANSI, *The U.S. Voluntary Standardization System*, 4-5.

54. ANSI, *The U.S. Voluntary Standardization System: Meeting the Global Challenge*, 18-22.

55. EEE, *Standards Catalog.*

56. NFPA, *National Electrical Code 1993.*

57. *ANSI Annual Report*, 1993, 13.

58. James Thomas, President, ASTM, personal communication, March 24, 1994; and Sergio Mazza, President, ANSI, personal communication, March 1, 1994.

59. ANSI, *The U.S. Voluntary Standardization System*, 18.

60. ANSI, *The U.S. Voluntary Standardization System*, 18.

61. Lehr, "Standardization: Understanding the Process," 551-552.

62. Stanley M. Besen and Joseph Farrell, *Choosing How to Compete: Strategies and Tactics in Standardization*, 117-131.

63. For further discussion of the interactions between standardization and technology development, see Farrell and Saloner, *Competition, Compatibility and Standards*; and Richard Jensen and Marie Thursby, *Patent Races, Product Standards, and International Competition.*

64. Cargill and Weiss, "Consortia in the Standards Development Process", 563-564.

65. Mary Anne Lawler, memorandum, U.S. Technical Advisory Group to JTC-1.

66. *ANSI Annual Report*, 1993, 8-9.

67. OTA, *Global Standards*, 48-49; 54-55.

68. Cargill, *Information Technology Standardization*, 142-147.

69. For descriptions of international standard-setting organizations and processes, see especially ANSI, *The U.S. Voluntary Standardization System*; Cargill, *Information Technology Standardization*, 125-148; and Stanley M. Besen and Joseph Farrell, *The Role of the ITU in Standardization*, 311-321. See also Maureen Breitenberg, ed., *Directory of International and Regional Organizations Conducting Standards-Related Activities*, NIST Special Publication 767.

70. ANSI, *The U.S. Voluntary Standardization System*, 27.

71. Data provided by ANSI, November 1993; and OTA, *Global Standards*, 87.

72. ANSI, *The U.S. Voluntary Standardization System*, 27-29; data on ISO/IEC committee participation provided by ANSI, July, 1994; and OTA, Global *Standards*, 87.

73. ANSI, *The U.S. Voluntary Standardization System*, 105-106.

74. ANSI, *The U.S. Voluntary Standardization System*, 30; and Michael Miller, President, Association for the Advancement of Medical Instrumentation, presentation to the Conference on New Developments in International Standards and Global Trade.

75. ASTM, internal memorandum, May 11, 1994.

76. ASME, "Annual Report for 1992/1993", AR-11.

77. For a directory of federal agencies with standards activities, see Toth, *Standards Activities of Organizations in the United States*, 547-656.

78. ANSI, *The U.S. Voluntary Standardization System: Meeting Global Challenges*, 14.

79. The regulation in question was a beer labeling requirement proposed by the State of

California. JoAnne Overman, Director, National Center for Standards and Certification Information, National Institute for Standards and Technology, personal communication, June 7, 1994.

80.   Breitenberg, *The ABC's of Standards-Related Activities in the United States.*

81.   NIST, *NIST Budget Summary*, August 30, 1994.

82.   Based on estimates provided by John Donaldson, Director, Standards Code and Information Program, Office of Standards Services, NIST.

83.   U.S. Department of Commerce, Memorandum, *Third Triennial Report to the Office of Management and Budget on the Implementation of OMB Circular A-119*, p. 15.

84.   James E. Rountree, *Directory of DOC Staff Memberships on Outside Standards Committees*, 3-6.

85.   Unpublished data provided by JoAnne Overman, Director, National Center for Standards and Certification Information, NIST, 1994.

86.   *The Office of Standards Services*, brochure published by NIST, n.d.

87.   JoAnne Overman, *GATT Standards Code Activities of the National Institute of Standards and Technology 1992*, 3-4; and JoAnne Overman, personal communication, June 7, 1994.

88.   Information provided by NIST, Office of Standards Services staff, May and June, 1994.

89.   Office of Management and Budget, *Circular No. A-119*, 1982.

90.   OMB, Circular No. A-119, Revised, in *Federal Register*, October 26, 1993, p. 57644.

91.   OMB, Circular A-119, Revised, paragraph 7.a.(4).

92.   Increased government use of commercial product specifications in place of government-unique standards and specifications is one goal of federal acquisition reform legislation signed into law in 1994. See Title VIII, *Federal Acquisition Streamlining Act of 1993*, 103rd Cong., 2nd session, S. 1587.

93.   Cassell and Crosslin, *Benefits of the Defense Standardization Program*, A3-A5.

94.   Cassell and Crosslin, *Benefits of the Defense Standardization Program*, A6-A10.

95.   George Eads and Peter Reuter, *Designing Safer Products: Corporate Responses to Product Liability Law and Regulation*, 34-39.

96.   Eads and Reuter, *Designing Safer Products*, 112-115.

97.   For a discussion, see Cheit, *Setting Safety Standards*, 8-14 and 211-221.

98.   Eads and Reuter, *Designing Safer Products*, 106-115.

99.   Cheit, *Setting Safety Standards*, 224.

100.   OMB, Circular No. A-119, Revised.

101.   Interagency Committee on Standards Policy.

102.   U.S. Department of Commerce, *Third Triennial Report to OMB*, p. 3.

103.   U.S. Department of Commerce, *Third Triennial Report to OMB*, p. 4.

104.   OTA, *Global Standards*, 46-58.

105.   *Trade Agreements Act of 1979*, 96th Cong., 1st sess., H.R. 4537.

106.   ANSI, *1993 Annual Report*, 24.

107.   Membership dues provide for meeting 43 percent of ANSI's core (non-publishing) expenses, primarily administration of U.S. consensus standards development. Net income from publications sales provides for 34 percent of these expenses; program support, 14 percent; conformity assessment services, 4 percent; and other, 5 percent. ANSI, *1993 Annual Report*, 13; and Manuel Peralta, President, ANSI, personal communication, October 25, 1993.

3

Conformity Assessment

In Chapter 1, seven functions of product and process standards were described: fostering commercial communication; diffusing technology; raising productive efficiency; enhancing market competition; ensuring physical and functional compatibility; improving process management; and enhancing public welfare (see Table 1-1). To succeed in these functions, standards must be well designed, based in sound technology, appropriate to the task at hand, and accepted as valid and useful by the population of users. The U.S. system for developing standards that meet these conditions is examined in Chapter 2.

A standard that meets these criteria, however, still fails to have the effect its developers intended if products designed to conform to it do not, in practice, conform. *Conformity assessment is the comprehensive term for measures taken by manufacturers, their customers, regulatory authorities, and independent, third parties to assess conformity to standards.* Conformity assessment and standardization are separate activities. The two are, however, closely related. Conformity assessment depends on the existence of unambiguous standards against which products, processes, and services are assessed. Conformity assessment enhances the value of standards by increasing the confidence of buyers, users, and regulators that products actually conform to claimed standards.[1]

The United States has an extensive and increasingly complex conformity assessment system.[2] Like the standards development system, it has evolved in a decentralized manner. As the needs of industry, government and society have changed and grown, particularly in the past 20 years, new elements and new layers of complexity have become part of the system. While each element has been motivated by specific marketplace or regulatory demands, the overall growth of the system has been uncoordinated. As discussed in this chapter, the result is the imposition of large costs associated with duplication, redundancy, and unnecessary complexity on testing laboratories, product certifiers, manufacturers, and ultimately, their customers.

Conformity assessment comprises four areas (see Figure 3-1). For convenience of discussion, the terms used in this figure focus on manufactured products.[3] The same concepts, however, also apply to conformity of processes and services. The first area, *manufacturer's declaration of conformity*, is assessment by the manufacturer based on internal testing and quality assurance mechanisms. Second is *testing* of products, parts, and materials performed by independent laboratories as a service to the manufacturer. Independent testing may be of value to the manufacturer as an outside confirmation of in-house test results; it may be required by a customer as a condition of sale; or it may be mandated by a regulatory agency. Independent testing services may also enable small manufacturing firms to operate without the need to maintain an in-house testing capacity. The third area is *certification*, formal verification by an unbiased third party, through testing and other means, that a product conforms to specific standards. Familiar examples of certification, among many others, are the Underwriters Laboratories product safety certification (the UL mark) and the U.S. Department of Agriculture quality grade for meat and poultry. The final area is *quality system registration*, the result of independent audit and approval of the manufacturer's quality system. A quality system is a management system, including procedures, training, and documentation, for ensuring consistency in product quality. Quality system registration is not an assurance of product quality; rather, it is a component of broader mechanisms for assessing products.

The purpose of these conformity assessment activities is to provide the relevant parties—such as the purchaser of a product or the regulatory agency with authority over a product—with whatever degree of confidence is needed in a particular circumstance. For a purchaser, that circumstance is the decision on whether to buy; for a regulator, it is the decision to approve or disallow the product for use or installation. In the absence of independent assurance of product conformity to standards, a purchaser or regulator must take the manufacturer's word that the product conforms. In most situations, as discussed below, this level of assurance—the manufacturer's declaration of conformity—is entirely sufficient and appropriate. Other elements of the conformity assessment system have evolved to meet the need for additional assurance in specific situations. The uncoordinated manner in which the system has grown and continues to grow, however, has raised costs and created obstacles to both domestic and international commerce.

This chapter identifies strengths and weaknesses in the U.S. conformity assessment system. To the extent possible, given the limited availability of economic data about the system, the economic impact of these inefficiencies is also examined. Interconnections between U.S. and international conformity as-

	Manufacturer's Declaration of Conformity	THIRD – PARTY ACTIVITIES		
		Product Testing	Product Certification	Manufacturing Processes: Quality System Registration
LEVEL 1: ASSESSMENT	Manufacturer's own testing and quality assurance BY: manufacturer	Testing of products, components, materials, etc. BY: independent laboratory	Certification of products against a standard or set of standards BY: product certifier	Audit and registration of manufacturer's quality assurance system (e.g., against ISO 9000 standards) BY: quality system registrar
LEVEL 2: ACCREDITATION	Acceptance BY: customer or regulatory authority	Accreditation of laboratory's competence BY: laboratory accreditation program (private or government)	Accreditation of certifier BY: certifier accreditation program (private or government)	Accreditation of quality system registrar BY: registrar accreditation program (private or government)
LEVEL 3: RECOGNITION	Acceptance BY: customer or regulatory authority	Official recognition of laboratory accreditation program BY: government *	Official recognition of certifier accreditation program BY: government *	Official recognition of registrar accreditation program BY: government *

FIGURE 3-1 Conformity Assessment System Framework (mechanisms for ensuring that products conform to standards).

NOTE: ISO = International Organization for Standardization.

* Government recognition programs are in very early stages of development.

sessment systems and their effect on U.S. international trade performance are discussed in Chapter 4.

CONFORMITY ASSESSMENT SYSTEM FRAMEWORK

The first level of the framework shown in Figure 3-1, *assessment*, represents the primary level at which the four activities of conformity assessment take place. At level 1, manufacturers, testers, certifiers, and quality system registrars evaluate products, processes, and services. These evaluations are the direct substance of conformity assessment—the comparison of a product to a standard. The second and third levels of the framework, *accreditation* and *recognition*, represent activities to evaluate the competence of the assessors operating at levels 1 and 2, respectively. Accreditation (of laboratories, certifiers, and registrars) and recognition (of accreditors) add additional layers of complexity and expense to the system. They have evolved in response to specific commercial and public-sector demands. They add cost to the system, however, and have frequently been implemented in an uncoordinated, redundant fashion. This is the case in both the public and the private sectors, discussed below.

Manufacturers' internal assessment procedures, leading to a manufacturer's declaration of conformity, are the simplest and oldest form of conformity assessment. The vast majority of commercial transactions take place without third-party assessment of product conformity.[4] Most manufacturers, especially large- and medium-sized firms, conduct their own testing and quality assurance to some degree of precision. In the marketplace, the buyer looks at a product, reads the packaging or advertising, and makes a decision. The buyer accepts the manufacturer's statements about the features of a product based on trust—and, to the extent feasible, on inspection of the product before buying it. This trust is, of course, founded on the customer's freedom to switch to a competitor's product if dissatisfied. Success and failure in the marketplace give manufacturers powerful incentives to support claims and maintain consistent quality. Truth-in-advertising laws, which are enforced by the Federal Trade Commission, also motivate manufacturers to ensure that products conform to advertised characteristics.[5]

The threat of private liability claims related to nonconforming products is also a strong incentive facing manufacturers. Conformity to safety standards—whether voluntary or regulatory—does not necessarily protect manufacturers against damage awards in product liability lawsuits.[6] A finding that a product that harmed someone failed to conform to relevant standards, however, is highly likely to result in an award of compensation. It is therefore in the manufacturer's interest to verify compliance with relevant standards. Certification of conformity by a third party does not free the manufacturer from liability. As a result, manufacturers of potentially dangerous products generally maintain internal testing and assurance procedures, even if they also seek third-party assessment.[7]

In some situations, a purchaser needs a stronger guarantee of product conformity than is provided by the freedom to change suppliers. This is increasingly true in capital-goods sectors. A case in point is that of manufacturers who purchase large volumes of parts, systems, or materials from suppliers (who, in turn, may purchase from lower-tier suppliers). The purchase contract between manufacturer and supplier provides formal specifications, or standards, that the supplied products must meet. If the purchaser must wait until delivery to inspect the product and send defective parts back to the supplier, the cost of delays in production while waiting for replacement parts—or finding a new supplier—can be very high. The rise of "just-in-time" manufacturing processes, with mini-mized inventories and requirements for instant supply of defect-free parts from external suppliers, has increased the cost of delays.[8]

Second-party conformity assessment is an outgrowth of this type of demand for assurance. In second-party assessment, only the two parties, supplier and purchaser, are involved, and the purchaser's own inspectors perform the assess-ment. By inspecting the supplier's production line, manufacturing processes, and samples or batches of parts before they leave the supplier's factory, a purchaser can gain confidence in the supplied products and reduce the potential for delays in his or her own production line. The benefit of obtaining this assurance, however, must be weighed against the cost of performing assessments.[9]

Third-party assessment is the sector of the conformity assessment system that has grown most in recent years.[10] In most commercial interactions, there is no need for the added expense and complexity of third-party conformity assess-ment. There are, however, two sets of circumstances in which relying on the manufacturer's declaration and the purchaser's own assessment is inadequate. In these situations, assessment by a neutral third party is necessary or desirable. First, concerns about the safety, health, or environmental impact of a product are sometimes too important to be left to the manufacturer's own assessment and too expensive or technically difficult for the customer to perform. This is true, for example, of products whose failure could lead to injury, illness, property damage, or loss of life. In these cases, it is unacceptable to discover the product's noncon-formity after a failure has occurred.

Much of the U.S. conformity assessment system exists specifically to ad-dress this type of safety, health, and environmental concern. In regulated product sectors, such as aircraft, automobiles, agricultural chemicals, heavy machinery, and drugs, a regulatory authority requires competent, prior assurance of confor-mity to relevant standards before a product can be accepted and used.[11] At lower levels of risk, this assurance may simply be the manufacturer's own declaration of conformity. This level of assurance imposes the least costs on industry and consumers. Many regulations, however, require third-party assessment to verify product safety. Drug safety certification required by the Food and Drug Admin-istration is an example of a federal program of this type.[12] Third-party assess-ment requirements for regulatory enforcement should be limited, ideally, to product sectors in which serious risk of harm justifies the cost burden of imposing third-party assessment.

Unregulated products may also be subject to third-party assessment as a result of marketplace demands. Purchasers may choose to demand independent, private assessment of product safety. They may, for example, buy only products bearing a recognized third-party certification mark, such as the Underwriters Laboratories UL mark or the American Gas Association Laboratories seal. One major retail chain, as a matter of policy, rarely markets electrical appliances not bearing the UL label.[13] Many voluntary product safety standards, and the institutional mechanisms that assess conformity to them, were in fact developed to meet needs not covered by government regulations. In some cases, private certification programs are developed by an industry to forestall government regulatory intervention.[14]

The second category of demand for independent assessment applies primarily to the relationship between manufacturers and their primary, secondary, and tertiary suppliers of parts and materials. As noted previously, purchasers may demand prior assurance that parts will conform to contract specifications, rather than relying on postdelivery inspection. The purchaser may choose to rely on a neutral third party to provide this assurance, rather than performing these assessments directly. Whether assessment is second party or third party, a range of approaches is available. For completeness, every single part could be inspected for conformity to the contract specification. Inspection of 100 percent of the parts is costly, however, and justifiable only if the consequences of a single nonconformity would be severe. An intermediate form of assessment involves inspection of samples of the supplier's product, combined with an assessment of the supplier's overall system for maintaining consistent product quality.

In select circumstances, third-party conformity assessment has advantages over second party for meeting the needs of suppliers and manufacturers. In an industry in which each supplier sells to many purchasers, it is redundant for each supplier to be audited and approved by every manufacturer, all performing essentially the same assessment. A single assessment of the supplier by a competent third party can, in this case, replace multiple second-party assessments.[15]

The remainder of this chapter focuses on third-party conformity assessment, rather than second party or manufacturer's declaration. There are two reasons for this focus. First, the activities of a manufacturer and purchaser to assess conformity are less a matter of public policy than they are the internal, competitive concern of the firm. They are intimately tied to research and development, testing, manufacturing process management, inventory control, and other activities within a firm's operations. As such, they are not directly within the scope of this study. Second, third-party assessment is the portion of the system in which the greatest growth and complexity have appeared, as discussed in the next two sections.

TESTING AND CERTIFICATION

Product Testing

Independent laboratories that perform testing services for clients comprise the largest share of the U.S. conformity assessment system. The Bureau of the Census publishes an annual survey of U.S. service industries. This survey provides an indication of the size and rapid rate of growth of the independent testing industry. Testing laboratories (Standard Industrial Classification Code 8734) that are subject to federal taxes accounted for a total revenue of more than $5.1 billion in 1992, an increase of 6.8 percent over 1991 (see Table 3-1). An average annual growth of 13.5 percent from 1985 to 1992 indicates a very high rate of expansion in this sector, mirroring the overall rapid growth in third-party conformity assessment in the United States.[16] More detailed data are available in the

TABLE 3-1 — U.S. Independent Laboratory Testing Services — Total Revenues (millions of dollars)

YEAR	TOTAL [a]	FOR-PROFIT FIRMS	NOT-FOR-PROFIT FIRMS [a]
1985	2,324	2,121	203
1986	2,370	2,163	207
1987	2,624	2,395	229
1988	3,438	3,138	300
1989	4,180	3,815	365
1990	4,977	4,543	434
1991	5,330	4,817	460
1992	5,637	5,145	492

[a] Estimated, except 1987.

SOURCES: U.S. Bureau of the Census. *Census of Service Industries: Subject Series, Miscellaneous Subjects*. Summary pp. 4-9 and 4-11. Washington, D.C.: U.S. Government Printing Office, 1987.

U.S. Bureau of the Census. *Service Annual Survey: 1992*. P. 18. Washington, D.C.: U.S. Government Printing Office, 1994.

U.S. Bureau of the Census. *Service Annual Survey: 1991*. P. 25. Washington, D.C.: U.S. Government Printing Office, 1993.

Census of Service Industries, published every five years. The most recent edition, for 1987, divides revenues in the industry between firms subject to federal income tax (91 percent of the total) and tax-exempt, or not-for-profit, institutions (9 percent).[17]

These data do not, however, capture the full scale of third-party testing. Many of the more than 400 members of the American Council of Independent Laboratories (ACIL), the industry association for testing laboratories, are classified as *engineering services* firms rather than laboratories.[18] These firms are members of ACIL because a significant share of their business consists of testing services. Their revenues from testing are not, however, separately identified in the Census Bureau data.[19] Engineering services totaled $61.5 billion in revenue in 1992.[20] When the number and size of ACIL members that identify themselves as engineering services firms are taken into account, and the Census data are scaled accordingly, independent laboratory services in the United States are estimated to be a *$10.5 billion industry.*

Testing services encompass a broad spectrum of technical activities and competencies. The International Organization for Standardization (ISO) definition of a test, in the context of conformity assessment, is a "technical operation that consists of the determination of one or more characteristics of a given product, process or service according to a specified procedure."[21] Materials, parts, and completed products may all be tested for their physical properties, such as strength and durability; physical dimensions; electrical characteristics, including interference with other electrical devices; acoustical properties; chemical composition; presence of toxic contaminants; and multitudes of other features.

Testing laboratories serve several categories of clients. The 1987 Census of Service Industries separates the receipts of independent testing laboratories, by class of client, into 8.7 percent federal government, 4.9 percent state and local governments, and 86.4 percent other clients—mainly private industry. Manufacturers rely on independent testing as a check against their own tests. Testing against specific standards provides independent data to support manufacturer's declarations of conformity to purchaser specifications or government regulations. Purchasers—including large manufacturers and government procurement agencies—rely on third-party testing to verify the conformity of parts and products to their stated specifications. Regulations frequently require manufacturers to show compliance through results of independent testing. The Occupational Safety and Health Administration (OSHA), for example, requires equipment used in the workplace to be tested by an independent laboratory accredited under OSHA's Nationally Recognized Testing Laboratory (NRTL) program.[22]

Product Certification

Certification is a form of conformity assessment that involves determining whether a product, process, or service meets a specific standard or set of standards. Certification is a "procedure by which a third party gives written assurance that a product, process or service conforms to specified requirements."[23] It is, by definition, exclusively a third-party activity. In the past, the manufacturer's declaration of conformity was sometimes referred to as "self-certification." The term caused confusion, however, and has been dropped both internationally and by the American National Standards Institute (ANSI) accredited committee for writing certification procedural standards, Committee Z34.[24]

Certification usually requires performance of product tests. The testing component of U.S. certification activities is included in the industry revenues data in Table 3-1; however, aggregate data on revenues from certification as a whole are not available. Certification is distinguished from testing by three key features. Certification always measures a product (or process or service) against one or more specific standards, whether mandatory, voluntary, or de facto. Testing, by contrast, does not necessarily measure against any specific standard. Second, certification is always performed by a third party, independent of either the supplier or the purchaser. Finally, certification results in a formal statement of conformity—a certificate—that can be used by the manufacturer to show compliance with regulations, meet purchasing specifications, and enhance the product's marketability. The certifier often licenses the manufacturer to print a *certification mark* on the product or its packaging, potentially increasing its acceptability to the buying public. Certification marks are the property of the certifier and are registered with the U.S. Patent and Trademark Office.[25]

Certification may encompass many different levels of complexity and expense, depending on the characteristics of the product and the degree of need for confidence in the product's conformity to standards. The more complex and intrusive the certification program is, the greater is its cost.[26] In sectors with high demands for safety and reliability, certifiers may require a relatively intensive certification process, involving multiple tests, one or more factory inspections, and testing of large numbers of product samples. Lower levels of need for assurance may be satisfied by type testing—the testing of one or a few samples as typical of all products with the same design and materials. Some certification programs require follow-up testing of additional samples obtained at the factory or on the open market in order to maintain certified status. Evaluation of the manufacturer's quality assurance system is part of some certification schemes, as discussed in the next section.

Private and Public Certification Programs in the United States

There are more than 110 private-sector product certifiers in the United States.[27] Many private-sector certifiers are also independent testing laboratories. Some certifiers, mainly those operating smaller programs, certify products on the basis of tests performed by other facilities. These tests must be performed by laboratories that are independent of the manufacturer. Whether testing is performed by the certifier or an independent laboratory, the certifier's role is to interpret the standard and judge whether the test results justify declaring the product to be in conformance.[28]

The majority of third-party certifiers in the United States are private, for-profit testing laboratories. As discussed in the previous section, these represent a large and growing service industry. In addition to providing testing services, many of these laboratories take the additional step of certifying products as meeting particular standards. Members of the ACIL that test and certify products include, among many others, ETL Testing Laboratories, for consumer appliances, sports equipment, safety glass, and other areas; United States Testing Company, for areas such as toy safety and toxicology; and MET Electrical Testing Company, for workplace safety, telecommunications equipment, and others.[29]

A number of broadly familiar certification programs, many of which incorporate their own certification marks, are conducted by private, not-for-profit organizations. Underwriters Laboratories (UL), founded in 1894, is one of the oldest certifiers in this country. UL is a major standards developer in the consumer product safety area, with more than 600 published safety standards.[30] It is also a leading tester and certifier of products, devices, and materials. UL certification of product safety—known as "listing" the product—authorizes the manufacturer to print UL's certification mark on the product or its packaging. Another not-for-profit testing and certification organization is the Factory Mutual Research Corporation. Factory Mutual tests and lists approved products as part of a series of activities to reduce industrial property damage.[31]

NSF International is a private, not-for-profit certifier in the areas of public health and the environment. NSF, like UL, is also a developer of ANSI-approved consensus standards. NSF's product standards and certification activities include, for example, a drinking water additives program initiated in 1985 under a cooperative agreement with the U.S. Environmental Protection Agency (EPA). NSF certifies products for compliance with ANSI/NSF Standards 60 and 61, for drinking water treatment chemicals and water system components. These certifications are accepted for regulatory purposes by the EPA and state regulators.[32]

The American Gas Association (AGA) certification program has been in operation since 1925. AGA tests in its own laboratories and certifies gas appliances and accessories, including furnaces and cooking appliances. Requirements for AGA certification, besides testing, include a review of design information and construction parameters, as well as factory and quality control inspections.[33] Other industry association-operated certification programs include those of the Air Conditioning and Refrigeration Institute, for air conditioners and water coolers, and the Association of Home Appliance Manufacturers, for refrigerators, air conditioners, and dehumidifiers.

The National Board of Boiler and Pressure Vessel Inspectors certifies boilers and components, water heaters, and nuclear reactor installations for compliance with the American Society of Mechanical Engineers' Boiler and Pressure Vessel Code. A professional association-operated program that is likely familiar to many consumers is that of the American Dental Association (ADA), whose certification mark is printed on toothpaste tubes. The ADA assesses product specifications provided by the manufacturer and tests samples purchased on the open market.[34]

Construction and building materials certification is an area of great activity and complexity in the U.S. system that overlaps the public and private sectors. As noted in the previous chapter, state and local governments are responsible for establishing safe building codes. To meet this responsibility, hundreds of these agencies have established mandatory requirements by reference to private certification programs. Private programs are operated by a variety of model building code organizations. These include the Building Officials and Code Administrators International, the International Conference of Building Officials, and the Southern Building Code Congress International.[35] These organizations compete with one another for certification business. In the absence of reciprocal recognition among these programs, manufacturers of building products and materials must seek multiple, redundant certifications to sell in multiple jurisdictions.[36]

States are active in other areas of product certification in addition to building materials. The most recent comprehensive directory of such programs, compiled by NIST in 1987, identifies states with regulations in products sectors ranging from agriculture and alcoholic beverages to consumer goods, machinery, and transportation.[37] Some of the sectors with the broadest coverage among states are agricultural commodities, regulated by 43 states; plant nursery stock, 47 states; road and bridge construction materials, 49 states; and measuring and weighing devices, all 50 states.

Requirements for certification in most product sectors vary by state. Activities to harmonize these requirements on a nationwide basis, however, are limited. One exception is in the area of weighing and measuring devices. These devices are required to be certified in order to ensure accurate, reliable measurement of commodities for sale. Under the leadership of the National Institute of Standards and Technology (NIST), the National Conference on Weights and Measures has worked to foster mutual recognition among more than 3,000 state and local weights and measures authorities throughout the United States. Through a network of agreements among these authorities, products may be weighed, measured, and packaged in one jurisdiction without having to be remeasured when shipped elsewhere in the United States.[38] Although specific estimates of U.S. economic benefit from this program do not exist, it is clear that the economies of scale created by this country's large domestic market would be reduced if the free flow of domestic commerce were interrupted for reweighing of packaged products at state lines.

A 1988 directory published by NIST lists 84 certification programs run by federal agencies. These draw their authority from a range of federal laws and regulations. Public sector certification programs serve three principal functions.[39] The first is to create a level playing field for commerce by assessing and enforcing standards for the quality of products for sale. An example is U.S. Department of Agriculture (USDA) certification of meat and poultry quality. Certification against USDA standards is voluntary, but the marketing advantage it provides is sufficient incentive for many producers to participate.[40] This category of activity relates closely to the federal role in maintaining reference standards as a public service, as discussed in Chapter 2.

A second set of federal programs certifies products against health, safety, and environmental regulations. Not all federal regulatory standards are accompanied by third-party certification requirements. In many industries, regulators accept the manufacturer's declaration of conformity to regulations. The automotive industry, in which the National Highway and Traffic Safety Administration (NHTSA) accepts manufacturers' declarations, is one example. Independent testing is generally performed by NHTSA only in the event of an actual or suspected product failure.[41] Numerous other federal certification programs exist in connection with the broad mandates of regulatory agencies, such as the Federal Aviation Administration, for aircraft components, and the Food and Drug Administration, for pharmaceutical products.[42]

The third category of federal certification programs concerns product testing for public procurement. The Department of Defense (DoD) is the main agency conducting this type of activity. For example, DoD requires manufacturers to show conformity to military specifications through independent testing, resulting in DoD certification. Through the Qualified Products List (QPL) program, DoD also conducts its own testing and certification of parts, materials, and products in order to guarantee their quality for military use. The QPL program also has the aim of streamlining the procurement process, by eliminating the need for manufacturers to recertify products for each separate purchase.[43]

No currently available data indicate the total cost to manufacturers and their customers of federal, state, and local certification mandates associated with regulatory and procurement standards. Apart from several directories of public and private programs compiled by NIST at several-year intervals, no comprehensive source of information about certification exists.[44] Determinations to impose regulatory and procurement certification decisions, which add layers of conformity assessment to affected manufacturing processes and commercial transactions, should be weighed against the benefits to the public interest of added assurance. The lack of data on which to base these decisions is, accordingly, cause for concern.

The proliferation of programs in the federal, state, local government, and private sectors suggests strongly, nevertheless, that significant savings could be achieved through consolidating U.S. certification programs. In a $10.5 billion independent testing industry, plus an as-yet-unmeasured level of expenditure linked to manufacturer's internal testing against certification criteria, streamlining could potentially achieve savings to industry and to consumers measured in the billions of dollars. Streamlining could be accomplished through such means as mutual recognition of equivalent certification programs; harmonization of programs to achieve equivalence; privatization of public certification programs; and central coordination of federal certification policy. These options are discussed in detail in the section on accreditation in this chapter.

QUALITY SYSTEM REGISTRATION

The most recent element in the growth of the U.S. conformity assessment system is the rapid spread of quality system registration. This trend has been stimulated by European Union regulations for product safety and by a growing market demand—particularly, outside the United States—for independent assessment of producers' quality management systems.[45] The best-known and fastest-growing aspect of this trend is registration to the ISO 9000 standards, a series of quality system standards published in 1987. Adoption of these standards by the Commission of the European Community in 1989, as part of its Global Approach to Testing and Certification, significantly accelerated their use worldwide, including in the United States. The number of U.S. firms registered to ISO 9000 has grown rapidly, from 279 at the end of the first quarter of 1992, and 1,259 in the first quarter of 1993, to 3,165 in the first quarter of 1994.[46]

Quality system registration is the assessment and periodic follow-up audit of a manufacturer's quality assurance system. Assessment and audit are performed by an independent party, the quality system registrar. The concept of quality embodied in modern quality assurance originated with Dr. W. Edwards Deming. Following the Deming model, a quality assurance system is a production management tool for monitoring and controlling variables in the manufacturing process that introduce variability and lead to defects. The system comprises elements such as documentation, training, statistical monitoring of results, and continuous improvement.[47]

Awareness of quality system registration has expanded rapidly in recent years, in conjunction with global growth in demand for the ISO 9000 series of standards. ISO 9000, like other quality system documentation standards, was designed primarily as an instrument for supporting high-volume transactions, such as those between manufacturers and their parts and materials suppliers. There is danger that much of this demand is due to misunderstanding of the ISO 9000 series' link to product quality—in some instances through overpromotion by the burgeoning service industry of quality system auditors, registrars, trainers, and consultants.[48] Despite its rising popularity in European consumer markets, it was not intended, and is not appropriate, as a certificate of product quality for use in the consumer marketplace. It is the total quality of the product, including its design and features as well as its freedom from defects, that is of concern to customers.[49]

Registration evaluates the quality assurance system, *not* the quality of the products themselves. The purpose of third-party quality system registration is to support the manufacturer's claim that a system is capable of delivering a product of consistent quality with little or no variation. Product quality is the result of two factors: consistent manufacturing processes and good design. A product that conforms consistently, with zero defects, to a poorly designed standard is not a product that most customers would consider high quality. Such a product could conceivably, however, be produced by an ISO 9000-registered production facility. The ISO 9000 standards require the manufacturer to document production processes and develop procedures to improve them. The appropriateness of the product's design is outside the scope of the standards.[50] Furthermore, although there is the potential for documentation systems such as those prescribed by the ISO 9000 series to improve U.S. manufacturers' quality, other management tools such as Total Quality Management (TQM) or the guidelines for the Department of Commerce's Malcolm Baldrige National Quality Award are better suited to meeting those goals for many firms.[51]

Quality system registration to standards similar to ISO 9000 has a proper role in the U.S. manufacturing system, not as a standard for products, but as a conformity assessment tool. In specific, confidence in the supplier's quality system can simplify the certification process. This may be the case, for example, in the supplier-purchaser relationship. If the supplier has a sound quality system, once a sample of a product has been approved (by either the customer or a third party, such as a certifier), the customer can be confident that the supplier will be able to reproduce that success repeatedly with few or no defective units.[52] Quality assurance removes the need to inspect every item individually (100 percent testing). The producer's ability to deliver consistently identical parts—as assessed and verified by the quality system registrar—enables the purchaser to inspect a small subset of the delivered items and have confidence that the rest are identical to those that were inspected.[53]

In comparison to manufacturer's declaration and other, less intrusive forms of certification, third-party quality system registration is considerably more costly and is justified only by a strong need for independent assurance of conformity. As a replacement for 100 percent inspection and testing, however, a regime of batch testing combined with quality system registration can be less costly. For high-volume transactions requiring a high degree of confidence in product conformity, quality system registration may be appropriate. It has been applied in purchasing arrangements by government agencies (particularly the General Services Administration and the Department of Defense) and large manufacturers (such as automobile and aerospace firms).[54]

Quality system registration was applied successfully, for example, in a DoD pilot program for acquisition of electronic microcircuits. This program substituted assessment of manufacturers' processes for inspection of individual products. Approved manufacturers were placed on a Qualified Manufacturers List

(QML), allowing streamlined acquisition from those sources. A Defense Science Board study on the use of commercial components in military systems concluded that a QML-style program implemented throughout DoD could save $800 million annually in microcircuits alone.[55]

The DoD case, however, also illustrates a serious problem in the spread of quality system registration—the growing cost of multiple registrations. Regulators and procurers in DoD and other federal agencies, as well as major manufacturers in the private sector, are increasingly requiring suppliers to have their quality systems registered by third-party assessors. These requirements, however, frequently cite different quality system standards. The U.S. Nuclear Regulatory Commission (USNRC), for example, requires that suppliers of regulated components for nuclear power plants be registered to a quality system standard developed by the USNRC, 10 C.F.R. 50, Appendix B.[56] The DoD Qualified Manufacturers List is another, separate quality system standard, as are the ISO 9000 standards.

As a result, many producers now face mandates for registration to multiple— but substantially equivalent—quality system standards. For example, a producer of capital equipment could easily be faced with the cost of registration to a federal standard—such as DoD's QML program or the 10 C.F.R. 50 standard; to ISO 9000 for export to European markets; and to the specific quality requirements of major private customers, such as Ford Motor Company's Quality One standards.[57] Even in circumstances where registration is justified, multiple registration of a single producer's quality system is clearly redundant and wasteful.

The cost of registration can be substantial, even for large firms with quality systems already in place. For example, one large manufacturer, IBM Corporation, has estimated that registering its facilities to ISO 9000 in order to meet European market requirements cost the firm $100 million. Because IBM's facilities already had quality assurance systems in operation that met market demands, obtaining ISO 9000 registration yielded limited additional improvement in quality.[58]

In 1993, a survey of all North American firms registered to ISO 9000 found an average total cost of registration of more than $245,000 per firm. This excludes the additional cost of follow-up surveillance audits by registrars.[59] Although some survey respondents anticipated recouping their expenses through cost savings associated with improvement in their management systems, approximately half of respondents expected only limited savings and considered ISO 9000 registration a largely irrecoverable business expense.[60]

In summary, the proliferation of quality system registration requirements presents a serious burden to many U.S. manufacturers. To the extent that critical needs for independent quality assurance do exist in some circumstances, duplicative quality system registration requirements should clearly be eliminated. Consolidation of duplicative requirements will yield significant savings, both to U.S. industry and to federal regulatory and procurement agencies. Some initial progress has been made in this regard. In 1994, both DoD and the General Services Administration (GSA) adopted policies permitting procurement officers to accept ISO 9000 registration in lieu of equivalent military and federal quality system standards.[61] It is not yet clear, however, whether procurement officers will be required to accept such registration in every instance. In addition, federal regulatory agencies have not responded with similar changes in policy.

The growth of ISO 9000 and other quality system registration services worldwide has added substantial new cost and complexity to the conformity assessment framework illustrated in Figure 3-1. Work is now in progress to develop a set of international standards that have the potential, if similarly overused, to impose even greater waste and expense on U.S. firms. The ISO Technical Committee 207 is now developing environmental management system standards. U.S. industrial and government participants in the process are promoting the position that the committee should avoid a key flaw in the ISO 9000 standards—failure to include, within the standards, strict guidance to restrain inappropriate, excessive reliance on third-party registration by government and private parties.[62]

ACCREDITATION AND RECOGNITION

As this chapter has outlined, the U.S. conformity assessment system has expanded from declaration of conformity by the manufacturer to include third-party testing, certification, and quality system registration. These activities are illustrated in the first level of Figure 3-1. The system has grown, however, not only in the range of activities included, but also through the addition of new *layers* of assessment. The second level of the system framework is accreditation.

Accreditation Services

Accreditation is a procedure by which an independent party, the accreditor, evaluates and formally acknowledges the competence of a first-level conformity assessment body, such as a testing laboratory, product certifier, or quality system registrar. Accreditation activities arose in the United States in response to both private and public demands for assurance that conformity assessment organizations were competent.[63] Just as the cost of third-party activities such as certification must be weighed against the need for formal product assessment, the cost of accreditation systems must be weighed against their particular benefits.

The need for accreditation is greatest when the user—a purchaser or regulator—of a conformity assessment service is not able to assess for itself the competence of that service's provider. A trusted accreditor can assist users in selecting testing laboratories, certifiers, and registrars. This removes the need for manufacturers, for example, to conduct their own costly and time-consuming evaluations of independent laboratories.[64] Many large manufacturers require their suppliers' testing laboratories to be accredited as a condition for accepting suppliers' products. Rather than assess their suppliers' laboratories themselves, some manufacturers accept accreditation of these laboratories by a third party. In this case, achieving a single, third-party accreditation relieves suppliers of the burden of audits by each of their customers.

The National Aerospace and Defense Contractors Accreditation Program (NADCAP) is one example of a private assessment program established to reduce costs for an industry sector by streamlining the accreditation process. NADCAP was established by the Performance Review Institute, an affiliate of the Society of Automotive Engineers, to accredit laboratories and quality systems of suppliers to prime contractors in the aerospace and defense industry.[65]

Regulatory authorities rely on third-party accreditation to judge the validity of tests performed on such items as building materials and electrical equipment. For example, OSHA's NRTL program accredits laboratories as competent to test and certify products used in the workplace. Only certifications from accredited laboratories are accepted by OSHA as proving product compliance to its regulations.

Companies seeking registration of their quality assurance systems choose registrars, in part, on the basis of the registrars' accreditation. In the United States, the only nationwide system for accrediting ISO 9000 quality system registrars is operated by the Registrar Accreditation Board, under a joint venture with ANSI. For many independent laboratories and quality system registrars, third-party accreditation has become a basic requirement to attract customers.

Accreditation of a laboratory's or certifier's competence in a particular field typically involves review of the following elements, among others: technical procedures; staff qualifications; product sample handling; test equipment calibration and maintenance; quality control; independence; and financial stability. Teams of accreditors make on-site inspections of facilities and conduct interviews. A laboratory may be required to show its proficiency by measuring known test samples. To maintain accredited status, periodic reassessment, with follow-up testing and site visits, is required.[66]

In comparison to most of its trading partners, the United States has a decentralized system of accreditation. More than 100 public and private-sector accreditation programs are listed in Tables 3-2, 3-3, and 3-4. Accreditation programs exist at all levels of government and in many private organizations, such as trade and professional associations. Directories of laboratory accreditation programs compiled by NIST's Office of Standards Services, list 31 federal, 21 state, 11 local, and 40 private laboratory accreditation programs. These figures understate the number of state, local, and private programs, as a result of low survey response rates by these organizations during the compilation of directories.

The two largest programs in the United States both focus on testing laboratory accreditation. (Many certifiers are also testing laboratories, however, and are eligible for accreditation of their testing services.) The National Voluntary Laboratory Accreditation Program (NVLAP) is a fee-for-service program oper-

TABLE 3-2 — U.S. Federal Accreditation Programs

AGENCY	TITLE OF PROGRAM	PURPOSE
Department of Agriculture		
Animal and Plant Health Inspection Service	Accreditation of Lab Facilities for Contagious Equine Metritis	To approve labs to conduct diagnostic procedures for CEM, a highly transmissible venereal disease of horses
Federal Grain Inspection Service	Grain and Commodity Inspection and Weighing	To approve federal and state government and private-sector agents to inspect and weigh grains and test equipment and weights
Food Safety and Inspection Service	Accredited Laboratory Program	To accredit domestic nonfederal analytical chemistry laboratories to analyze meat and poultry food products
Office of Transportation	Agreement on the International Carriage of Perishable Foodstuffs and on the Special Equipment to Be Used for such Carriage (ATP)	ATP sets standards for testing and uses of equipment that carries perishable foodstuffs
Department of Commerce		
National Institute of Standards and Technology / Office of Weights and Measures	State Laboratory Program	To provide system for certifying state weights-and-measures laboratories by specific measurement areas
NIST / National Voluntary Laboratory Accreditation Program	National Voluntary Laboratory Accreditation Program	To accredit public and private testing laboratories
Department of Defense		
Army Corps of Engineers	CE District, Project and Commercial Laboratories Assurance Inspection Program	Each district requests assurance inspections of their testing laboratories by the division materials laboratory

Defense Electronics Supply Center	No title	To determine that commercial laboratories are equipped to perform testing for manufacturers listed on Qualified Product/Manufacturer Lists
Defense Logistics Agency / Defense Personnel Support Center	Qualified Laboratory List	To list laboratories capable of performing types of specification tests for clothing, textiles, footwear, and equipage-type items
Environmental Protection Agency		
Office of Drinking Water	Drinking Water Laboratory Certification Program	To evaluate laboratories that analyze public drinking water supplies for compliance purposes in states without primacy under the Safe Drinking Water Act
Office of Mobile Sources / Emission Control Technology Division	Retrofit Device Evaluation Program	To recognize independent laboratories performing preliminary screening tests on vehicles for assessing the emission and fuel economy of devices
Office of Air Quality Planning and Standards / Technical Support Division	Wood Stoves New Source Performance Standards Laboratory Accreditation	Manufacturers must use accredited laboratories to certify that their stoves meet air emission limits
Federal Communications Commission		
Office of Engineering and Technology / Authorization and Evaluation Division / Equipment Authorization Branch	Description of Measurement Facilities	Laboratories making measurements of devices that are included in applications for FCC equipment authorization provide a description of their facilities
Department of Health and Human Services		
Alcohol, Drug Abuse, and Mental Health Administration / National Institute on Drug Abuse / Division of Applied Research	National Laboratory Certification Program	To establish appropriate standards and procedures for periodic review of laboratories and criteria for laboratory certification for urine drug testing for federal agencies

TABLE 3-2 — (continued)

AGENCY	TITLE OF PROGRAM	PURPOSE
Food and Drug Administration / Bioresearch Program	Toxicology Laboratory Monitoring Program	To ensure the quality and integrity of data collected in order to support regulatory decision making
FDA / Center for Food Safety and Applied Nutrition	Evaluation of Milk Laboratories	To evaluate and endorse laboratory sample collection surveillance procedures to provide a national base for the uniform collection and examination of milk
Health Care Financing Administration / Health Standards and Quality Bureau	Clinical Laboratories Improvement Act of 1967 and Medicare	Licensure program required for laboratories in interstate commerce
Department of Housing and Urban Development		
Federal Housing Administration / Office of Architecture and Engineering / Materials Acceptance Division	Technical Suitability of Building Products Program	To accredit third-party organizations that validate manufacturers' certifications that building materials meet applicable requirements
Department of the Interior		
Office of Surface Mining Reclamation and Enforcement / Division of Technical Services	Small Operator Assistance Program	Laboratories prepare data for eligible small coal operators that are used for the operator's small coal mine permit application
Department of Labor		
Mine Safety and Health Administration / Approval and Certification Center	No title	Accreditation of inspection/test laboratories as a function of manufacturer quality control maintenance for component parts made and assembled for MSHA approval or certification

MSHA / A&CC	Testing and Evaluation of Illumination Systems	Evaluation to determine if light lab meets criteria
MSHA/ A&CC	Testing by Applicant or Third Party	Product testing at manufacturer or at third-party lab
Occupational Safety and Health Administration / Office of Variance Determination	Nationally Recognized Testing Laboratories Program	Applicant for recognition as a NRTL shall have administrative and technical capability to conduct product testing activities
OSHA / Blood Lead Program	Blood Lead Program	To evaluate proficiency testing data and approve labs that meet OSHA requirements and standards
OSHA / Office of Construction and Maritime Compliance Assistance	Maritime Cargo Gear Accreditation and Certification Program	Third-party providing impartial inspection of certain maritime cargo gear handling devices
Department of Transportation		
United States Coast Guard / Merchant Vessel Inspection Division	Lifesaving, Engineering, Fire Protection, and Pollution Prevention Equipment for Recreational Boats and Commercial Vessels	Laboratories test equipment to standards identified by the Coast Guard. Equipment is either certified by the lab or approved by the Coast Guard
USCG / Marine Technical and Hazardous Materials Division	Approval of Equipment for Use in Hazardous Areas Aboard Commercial Vessels	USCG requires electrical equipment in hazardous areas on certified vessels to be "listed" by an independent lab recognized by the Commandant
USCG / Merchant Vessel Inspection and Documentation Division	Delegated Approval Authorities for Safety Approval of Cargo Containers	To delegate to qualified persons or organizations authority to approve cargo containers used in international transport

TABLE 3-2 — (continued)

AGENCY	TITLE OF PROGRAM	PURPOSE
Department of the Treasury		
United States Customs Service / Office of Labs and Scientific Services	Petroleum Gaugers and Commercial Laboratory Programs	To accredit commercial labs to test petroleum, petroleum products, and organic chemicals in bulk and liquid form
Bureau of Alcohol, Tobacco and Firearms / Office of Laboratory Services	Certification of U.S. Laboratories for the Analysis of Wines and Distilled Spirits for Export	To certify labs qualified for analysis of wines and distilled spirits beverages to meet requirements of certain countries that require a chemical analysis
Department of Veterans Affairs		
Veterans Health Services and Research Administration / Pathology Service	College of American Pathologists Inspection and Accreditation Program and Related Quality Control Programs	Centralized contract through VA Marketing Center to provide quality control, inspection, and accreditation of clinical labs, special function, and nuclear medicine labs

SOURCE: Breitenberg, Maureen ed. *Directory of Federal Government Laboratory Accreditation/Designation Programs.* NIST Special Publication 808. U.S. Department of Commerce. Gaithersburg, Md.: NIST, 1991.

ated by NIST since 1976.[67] NVLAP accredits testing laboratory competence in a range of fields of testing. These include, among others, testing and analysis of asbestos fibers; construction materials; lighting, insulation, and paint products; carpeting; electromagnetic interference; measuring device calibration; personal radiation dosimetry; computer software compatibility; and fasteners. Accreditation fees vary by field, with an initial application fee ranging from $500 to $1,500; annual administrative and technical support fees from $2,600 to $5,600; and fees for on-site assessment of $1,500 to $2,300.[68] The 1994 NVLAP directory lists 670 accredited laboratories. The largest areas of accreditation, by number of laboratories, is asbestos analysis. Although NVLAP accreditation is voluntary in most fields, federal asbestos regulations, for example, require testing by a NVLAP-accredited laboratory.

The largest private-sector accrediting organization is the American Association for Laboratory Accreditation (A2LA). Established in 1986, A2LA accredited 603 laboratories as of January 31, 1994.[69] Mechanical, chemical, environmental, and construction materials are the predominant testing fields accredited by A2LA. Like NVLAP, A2LA accredits according to guidelines developed by the ISO Council Committee on Conformity Assessment for competence of testing laboratories. Accreditation fees charged by A2LA are typically somewhat lower than those of NVLAP, starting at $1,000 for a single field of testing, plus expenses for on-site inspection of the laboratory by independent assessors.[70]

Costs of Redundancy in U.S. Accreditation

The decentralized and complex nature of the U.S. system has arisen in an uncoordinated fashion through case-by-case response to specific demands such as those described above.[71] As a result, many of the programs listed in Tables 3-2 through 3-4 overlap. This imposes an unnecessary burden on laboratories and certifiers, which must obtain multiple accreditations for each of their areas of competence. For example, a laboratory seeking nationwide acceptance to conduct electrical safety-related materials testing must gain accreditation from at least 43 states; more than 100 local jurisdictions; 3 independent building code organizations; several federal agencies, including OSHA and NVLAP; and several large manufacturers. All of these accreditations evaluate the same laboratory for the same area of competence. This redundancy imposes unnecessary, unjustifiable costs on laboratories and their customers.[72]

A General Accounting Office study in 1989 identified an area of significant overlap in laboratory accreditation programs operated by the federal government. Although NVLAP operates a program in the field of electromagnetic interference—measurement of interference emitted by electronic devices—the Federal Communications Commission (FCC) does not recognize NVLAP accreditation. Instead, FCC requires laboratories that test products for compliance to its regulations to obtain a redundant accreditation from FCC.[73]

TABLE 3-3 — State and Local Government Accreditation Programs

AGENCY	TITLE OF PROGRAM	PURPOSE
Alabama		
Department of Environmental Management	Safe Drinking Water Act Certification	To require certification of all laboratories performing analyses under the Safe Drinking Water Act
California		
Department of Housing and Community Development	Approval or Listing of Testing and Inspection Agencies	Approval of standards plan for accessory structures or buildings, etc., that must be tested, inspected, certified by listed agency
State and Consumer Services Agency	Approval of Inspection and Labeling Agency	Approval of inspection and labeling agencies for California State Fire Marshal Listed Products
Connecticut		
State Department of Public Health Laboratory Approval and Registration Program	Connecticut State Department of Public Health Laboratory Approval and Registration Program	Approval and registration of private and municipal laboratories in the state as public health laboratories
Department of Public Safety	Approval of Testing Laboratories	For testing and listing of electrical equipment and apparatus and for gas-fired appliances
Florida		
Board of Professional Engineers / Department of Professional Regulations	Testing Laboratory Registration Rules	To register and regulate all activities of testing laboratories concerning engineering documents and reports so they meet standards of quality set forth in statutes and rules governing the practice of engineering

Agency	Program	Description
Office of State Fire Marshal	Listing as a Nationally Recognized Testing Laboratory	To accept products listed by nationally recognized testing laboratories meeting standards accepted by the Florida State Fire Marshal
State Solar Collector Certification Program	Florida State Solar Collector Certification Program	Approval of solar collector ratings as a mandatory program with laboratory accreditation of firms that provide necessary test services
Kentucky		
Cabinet for Natural Resources and Environmental Protection	Certification of Laboratories for Water Contaminants and Analysis	Accreditation of labs for inorganic and organic contaminant groups and microbiological analysis of public water supplies
Massachusetts		
Department of Public Safety / Board of Fire Prevention Regulations	Recognition for Testing Facilities and Listing of Electrical Equipment and Apparatus	Recognition of testing labs capable of testing and listing products called out in the applicable sections and provisions of the Massachusetts Electrical Code
State Building Code Commission Concrete Materials Testing Laboratory Accreditation	Accreditation of Concrete Materials Testing Laboratories	Licensing of all labs defined by the regulations and engaged in the testing of concrete or concrete materials for use in buildings and structures subject to control
New Hampshire		
Office of Fire Marshal, Electrician Board	Acceptance of Product Safety Testing	Acceptance of testing labs for the purpose of identifying product conforming to state code
New Jersey		
Department of Environmental Protection	Water Testing Laboratories Accreditation, Quality Assurance Program	Accreditation and licensing of laboratories that test water quality for residue and radiological contamination

TABLE 3-3 — (continued)

AGENCY	TITLE OF PROGRAM	PURPOSE
New York		
Health Department Environmental Laboratory Approval Program	Environmental Laboratory Approval Program	To approve labs to test samples from public water supplies where the data are to be used to show compliance with the state sanitary code
Department of State Office of Fire Prevention & Control	Fire Gas Toxicity Data File	Determination of testing agencies, acceptable to the Secretary of State, to perform combustion toxicity testing of building materials mandated by the state
North Carolina		
Department of Insurance	Acceptance of Laboratories Qualified to Provide Evidence of Safety of Electrical and Mechanical Goods	Acceptance of electrical products tested by laboratories accredited by the North Carolina Building Code Council
Ohio		
Board of Building Standards	Accreditation of Testing/Inspection Agencies	Accreditation of agencies for classification of materials or products and of labs for on-site and job-order construction testing of building construction material
Oregon		
Building Codes Agency	Testing Laboratory Accreditation	To establish labs whose tests, standards, and approvals are acceptable on electrical products
Pennsylvania		
Department of Agriculture / Division of Milk Sanitation	Division of Milk Sanitation	Approval of milk and dairy products testing facilities, sample collection, analysts' evaluation, and operations

Vermont		
Department of Labor and Industry	Electrical Testing and Labeling Laboratory Recognition	Recognition of testing laboratories to test and label products conforming to requirements of adopted codes and standards
Washington		
Department of Labor and Industries	Accreditation as an Electrical Testing Laboratory	Acceptance of labs as nationally recognized to approve controlled devices to meet the adopted code
LOCAL AGENCIES		
City of Los Angeles, Calif.		
Department of Building and Safety	Testing Agency Approval	Approval of all labs testing electrical and mechanical products whose published listings are recognized by the department
City of Oakland, Calif.		
Department of Public Works, Electrical Inspection	Approval of Electrical Safety Testing Laboratories	Approval of testing laboratories to label or list products as conforming to the requirements of the Superintendent of the Electrical Department
City and County of San Francisco, Calif.		
Bureau of Building Inspection	Approved Listing Agencies	Approval of listing agencies to test and list materials and methods of construction
Metropolitan Dade County, Fla.		
Product Control Section	Determination of Laboratories Acceptable	Approval of all products that are a permanent part of a building, tested to appropriate standards, and approval of labs that provide such data

TABLE 3-3 — (continued)

AGENCY	TITLE OF PROGRAM	PURPOSE
City of Atlanta, Ga.		
Department of Planning and Development	Code Compliance Office	To approve agencies that provide an identification symbol or label as a recognized testing laboratory
City of Chicago, Ill.		
Electrical Inspection Section	Accreditation as a Recognized, Independent Electrical Testing Laboratory	To accredit as nationally recognized standards testing laboratories, organizations whose testing and listing services are acceptable
City of New Orleans, La.		
Department of Safety and Permits	Recognized Safety Testing Laboratory	Evaluation of laboratories for recognition as safety testing laboratory
City of New York, N.Y.		
Department of Buildings	Acceptance of Testing Service or Laboratory	Determination of acceptability of testing and inspection laboratories and firms to provide data on compliance of products on installation

City of Cleveland, Oh.		
Board of Building Standards and Building Appeals	Approved Testing Laboratory Listing	To list approved testing laboratories for energy, electrical safety, and gas-fired appliances
City of Portland, Oreg.		
Bureau of Buildings	Electrical Testing Laboratory Approval Program	Approval of laboratories that are permitted to test and certify or label electrical equipment for sale or use in the city
City of Richmond, Va.		
Electrical Department	Laboratory Acceptance Program	Acceptance of laboratory judged to be a recognized testing lab for the inspection and approval of appliance, devices, or materials in accordance with the National Electrical Code

SOURCE: Hyer, Charles W. ed. *Directory of State and Local Government Laboratory Accreditation/Designation Programs*. NIST Special Publication 815. U.S. Department of Commerce. Gaithersburg, Md.: NIST, 1991.

TABLE 3-4 — U.S. Private Sector Testing Laboratory Accreditors

ORGANIZATION	FIELDS OF TESTING
Air Movement and Control Association	Acoustic/vibration measurement, construction materials, electrical, mechanical, and thermal
American Architectural Manufacturers Association	Construction materials, mechanical, and thermal
American Association for Laboratory Accreditation	Acoustic/vibration, biological, chemical, construction materials, electrical, geotechnical, mechanical, medical, metrology, nondestructive, optics and photometry, and thermal
American Association of Motor Vehicle Administration	Acoustic/vibration, chemical, electrical, mechanical, and optics and photometry
American Association of State Highway and Transportation Officials	Construction materials
American Institute of Steel Construction	Chemical, construction materials, and mechanical
American Society for Testing and Materials	Chemical and mechanical
ASTM and NIST Cement and Concrete Reference Laboratory	Construction materials
American Society of Mechanical Engineers	Biological, chemical, electrical, mechanical, metrology, and thermal
American Wood Preservers Bureau	Chemical, construction materials, and mechanical
Associated Laboratories, Inc.	Chemical and construction materials
Association of American Railroads	Acoustic/vibration measurement, chemical, mechanical, nondestructive
Board of Accreditation of Concrete Testing Laboratories, Inc. of North Carolina	Chemical, construction materials, mechanical
Building Officials and Code Administrations International	Chemical, construction materials, electrical, mechanical, nondestructive, and thermal
Cellulose Industry Standards Enforcement Program	Construction materials
Corporation for Open Systems International	Electrical, mechanical, and ionizing radiation
Council of American Building Officials / National Evaluation Service	Chemical, construction materials, electrical, mechanical, nondestructive, and thermal
ETL Testing Laboratories, Inc.	Mechanical and thermal
Insulating Glass Certification Council	Chemical and mechanical
International Association of Plumbing and Mechanical Officials	Chemical, construction materials, mechanical, and thermal
International Conference of Building Officials	Chemical, construction materials, electrical, mechanical, nondestructive, and thermal

TABLE 3-4 — (continued)

ORGANIZATION	FIELDS OF TESTING
International Electrotechnical Commission Quality Assessment System for Electronic Components	Acoustic/vibration measurement, electrical, mechanical, metrology, and thermal
Kitchen Cabinet Manufacturers Association	Mechanical
MET Electrical Testing Company, Inc.	Acoustic/vibration measurement, chemical, construction, electrical, mechanical, metrology, and thermal
MTL Certification Services Co., Inc.	Mechanical
National Association of Independent Laboratories for Protective Equipment Testing	Chemical, electrical, and mechanical
National Board of Boiler and Pressure Vessel Inspectors	Construction materials, mechanical, and thermal
National Certified Testing Laboratories	Mechanical and thermal
National Electrical Testing Association	Construction materials and electrical
National Environmental Balancing Bureau	Acoustic/vibration measurement, mechanical, and metrology
National Marine Manufacturers Association	Acoustic/vibration measurement, chemical, electrical, mechanical, and optics and photometry
National Safe Transit Association	Mechanical
National Wood Window and Door Association	Construction materials, mechanical, and thermal
Performance Review Institute / NADCAP	Acoustic/vibration, chemical, electrical, mechanical, metrology, nondestructive, optics and photometry, and thermal
Precast/Prestressed Concrete Institute	Construction materials
Safety Glazing Certification Council	Chemical and mechanical
Solar Rating and Certification Program	Construction materials, electrical, mechanical, optics and photometry, and thermal
Southern Building Code Congress International	Acoustic/vibration, chemical, construction materials, mechanical, nondestructive, and thermal
U.S. National Electronic Components Quality Assessment System	Acoustic/vibration measurement, electrical, mechanical, metrology, and thermal

SOURCE: Hyer, Charles W. ed. *Directory of Professional/Trade Organization Laboratory Accreditation/Designation Programs*. NIST Special Publication 831. U.S. Department of Commerce. Gaithersburg, Md.: NIST, 1992.

Another example of multiple accreditation requirements imposed at the federal level is associated with American Society for Testing and Materials (ASTM) standard testing methods in many product sectors. For example, the standard test method for radiant panels used in lighting is ASTM E648. NVLAP accredits labs for testing to E648. The NRTL program operated by OSHA, however, does not accept NVLAP accreditation in this area, requiring laboratories to obtain a second accreditation. In fact, the GSA Furniture Center; DoD's Defense Electronics Supply Center and Defense Logistics Agency; the U.S. Coast Guard; and the Federal Aviation Administration also require E648 accreditation. In addition, state and local governments may impose their own requirements and are under no obligation to accept any of these federal accreditations—with the exception of NRTL accreditation, which states are required to accept as a result of regulatory preemption by the Department of Labor. The ASTM standard for radiant panel testing is only one of many standard test methods subject to this degree of multiple accreditation.[74]

Data from three private, independent testing laboratories illustrate the cost burden of multiple accreditation in the U.S. system. Accreditation costs for these laboratories, all of which have average annual revenues from testing of less than $1 million, range from $12,900 to $87,000 per year.[75] To meet multiple accreditation requirements, one laboratory is accredited in a single area of testing, electromagnetic interference, by all of the following: NVLAP; the FCC; the U.S. Coast Guard Laboratory Approval Program; DoD's Defense Electronics Supply Center; and a European organization, Interference Technology International. The president of a second laboratory reports that accreditation costs "could double or triple within the next 3-5 years" because of "unnecessary duplicative costs."

Economic data on the aggregate costs to the U.S. economy of multiple accreditation in product testing are unavailable. A 1993 study in a related area, however, provides compelling evidence of potential savings from consolidating redundant accreditation programs. The EPA commissioned a study in 1992 by a special advisory Committee on National Accreditation of Environmental Laboratories (CNAEL).[76] Among other key findings, CNAEL concluded that environmental laboratories operating in multiple states face accreditation for the same tests by each state, often with arbitrarily differing criteria. CNAEL performed a detailed cost analysis and identified accreditation costs of $1,400 for small laboratories and between $10,773 and $21,546 for large laboratories. These costs include on-site audit costs, accrediting fees, and performance evaluation sample testing.

Aggregating these costs over the environmental testing industry, the CNAEL study found that replacing multiple state accreditation programs with a single, national program would significantly reduce costs. From a current, total cost estimated at between $17 million and $28 million per year, a streamlined system would reduce costs to between $13.5 million and $15.5 million, including a significant administrative fee charged by the national program.[77] This reduction of approximately 28 percent in national economic costs of accreditation would be highly significant if they could be replicated in the much larger sphere of product testing and certification. Accreditation costs and patterns of redundant accreditation requirements, of course, may vary considerably between environmental testing and the wide range of product testing fields. Nevertheless, the findings of the CNAEL study provide important guidance—particularly, in the absence of similar, detailed estimates for the product testing sphere.

Government Recognition of Accreditation Services

Recognition is the most recent layer to be added to the U.S. conformity assessment system. Recognition involves assessment of the competence of programs that accredit conformity assessment organizations, such as laboratories, certifiers, and quality system registrars.[78] Government recognition has the effect of conferring official acceptance, for example, of testing and certification performed by any laboratory accredited by a government-recognized accreditor. By relying on competent, private accreditation services to evaluate testing laboratories, instead of performing those evaluations directly, government agencies can reduce costs while continuing to meet their need for confidence in the reliability of product testing data.

Recognition programs, as indicated in Table 1-1, are in an early stage of development. Two U.S. government programs, both operated by NIST, currently involve recognition-level activities. These are the National Voluntary Conformity Assessment System Evaluation (NVCASE) program and NIST conformity assessment activities mandated by Congress under the Fastener Quality Act (P.L. 101-592).

The NVCASE program was established by NIST in early 1994. Its goal is to provide for increased access of U.S. products to foreign markets. NVCASE provides a basis for the U.S. government to give assurance to foreign governments of the technical competence of U.S. conformity assessment organizations. This assurance is intended to encourage foreign government acceptance of conformity assessment services performed by U.S. organizations—such as product testing and certification—as meeting foreign regulatory requirements.[79] In the absence of foreign government acceptance, U.S. products must be retested for conformity with product regulations in export markets, even when they have been tested and certified within the United States. Issues related to international acceptance of conformity assessment procedures are examined in detail in Chapter 4.

NVCASE is a voluntary program. U.S. conformity assessment organizations are not required to seek recognition. Once fully implemented, NVCASE will evaluate U.S. conformity assessment organizations at their request, on a fee-for-service basis, in accordance with internationally accepted standards for conformity assessment procedures.[80] NIST will list approved organizations in a register of recognized U.S. conformity assessment programs. NIST has stated that in order to limit expansion of the government's role in conformity assessment, NVCASE will operate principally at the level of *recognition of accreditors* and will recognize laboratories and product certifiers in a given product sector or field of testing only if no accreditation program exists in that area.[81]

In contrast to NVCASE, which applies only to exported products, NIST's activities under the Fastener Quality Act apply to conformity assessment associated with domestic regulatory requirements. The purpose of the act is to regulate the quality of fasteners for private commerce and public procurement.[82] The act requires NIST to establish programs both to accredit the competence of laboratories to test fasteners, and to *recognize* private accreditors of testing laboratories. This program creates a new precedent for U.S. government reliance on recognition of private accreditors to meet the needs of federal regulatory enforcement. As NIST identifies and recognizes private-sector accreditors, NIST-operated accreditation programs in this area will become duplicative and unnecessary. To the extent that recognized, competitive private-sector accreditation services take the place of government-operated programs, such as those under the Fastener Quality Act and others identified in Table 3-2, there is the potential for significant reduction in government costs.

Recognition-level activities are likely to increase in importance, for three reasons. First, government recognition has the potential to promote U.S. exports by enabling negotiation of government-to-government agreements for mutual recognition of conformity assessment systems. It was for this purpose that NVCASE was created. Prospects for mutual recognition agreement negotiations with U.S. trading partners, as well as the role of NVCASE in such agreements, are discussed in detail in the next chapter.

Second, recognition can support streamlining of the U.S. domestic conformity assessment system. Government recognition of private-sector accreditors will enable both regulatory and procurement agencies to eliminate the costs associated with operating federal accreditation, certification, and other conformity assessment programs while maintaining responsible oversight of regulatory and procurement enforcement. NIST recognition of private accreditors under the Fastener Quality Act provides a clear precedent for reliance on private-sector accreditors in regulatory enforcement. Transferring federal conformity assessment activities to the private sector will eliminate duplication between public and private programs and will increase the efficiency of the U.S. conformity assessment system. Specific mechanisms by which these goals may be achieved are elaborated in Chapter 5.

Finally, federal government recognition of private-sector conformity assessment programs has the strong potential to promote acceptance of conformity assessment among states and cities within the United States. As this chapter has noted, redundancy in testing and certification requirements is an acute problem at these levels. Federal recognition will increase the acceptability of conformity assessment procedures throughout the United States, reducing inefficiency and raising the level of competition in product testing and certification services.

SUMMARY AND CONCLUSIONS

This chapter has examined the growing size and complexity of the U.S. system for ensuring that products conform to standards. The procedures for carrying out this function are known collectively as *conformity assessment*. Product testing, certification, and laboratory accreditation services are key elements of the U.S. conformity assessment system. Testing and certification services provided by independent laboratories represent a $10.5 billion industry in the United States. Evaluation and registration of manufacturers' quality management systems is a new, rapidly expanding component of the system. A growing number of public and private programs independently accredit testing laboratories, certifiers, and quality system registrars.

This chapter documents serious waste and inefficiency in the U.S. conformity assessment system. The system's growing complexity, as well as the lack of coordination among federal, state, and local authorities, have serious, adverse consequences for U.S. economic performance. Measures must be taken in both the public and the private sectors to address these problems, including the gathering and analysis of additional data on the economic costs of redundant certification, registration, and accreditation, and other sources of inefficiency in the system. Raising the efficiency of the U.S. conformity assessment system will lower costs for private firms, government agencies, and consumers. It will increase the competitiveness of U.S. firms in both domestic and foreign markets. Chapter 5 presents specific recommendations to achieve these goals.

The next chapter places issues concerning standards development and conformity assessment systems into the context of international trade. It examines the increasingly close links between domestic and international standards, testing, and certification and the performance of U.S. manufacturers in global markets. Recent developments in U.S. multilateral and bilateral trade relationships are assessed, as well as U.S. policy measures that have the potential to improve U.S. export performance.

NOTES

1. International Organization for Standardization (ISO), *Certification and Related Activities*, 22-23.

2. Breitenberg, *The ABC's of Certification Activities in the United States*, 15.

3. For internationally accepted definitions of conformity assessment terms, see International Organization for Standardization, *Compendium of Conformity Assessment Documents*, 153-160. See also Breitenberg, *The ABC's of Certification Activities in the United States*.

4. ISO, *Certification and Related Activities*, 14.

    5.   See Breitenberg, *The ABC's of Certification Activities in the United States*, 2; and Robert
B. Toth, ed., *Standards Activities of Organizations in the United States*, 597-598.
    6.   For a discussion stressing the impact of product liability on corporate decisions regarding
product safety, see George Eads and Peter Reuter, *Designing Safer Products*.
    7.   ISO, *Certification and Related Activities*, 141-145.
    8.   For an overview of the evolution of the U.S. conformity assessment system framework,
and its link to changes in manufacturing and business practices, see Locke, *Conformity Assessment—
At What Level?*
    9.   For discussion, see ISO, *Certification and Related Activities*, 41-50.
    10.   For discussion, see Locke, *Conformity Assessment—At What Level?*
    11.   For an overview of areas regulated through federal conformity assessment programs, see
Breitenberg, ed., *Directory of Federal Government Certification Programs*.
    12.   Breitenberg, *Directory of Federal Government Certification Programs*, 145-148.
    13.   Cheit, *Setting Safety Standards*, 9.
    14.   Eads and Reuter, *Designing Safer Products*, 41.
    15.   For a discussion of circumstances in which third-party conformity assessment in place of
second-party is justified, Keith Mowry, *Conformity Assessment: An Extra Benefit from Standards*.
    16.   U.S. Bureau of the Census, Current Business Reports BS/92, *Service Annual Survey:
1992*, 18.
    17.   U.S. Bureau of the Census, *1987 Census of Service Industries: Miscellaneous Subjects*,
Summary 4-9, 4-11.
    18.   In 1994, ACIL changed its name to reflect more accurately the range of services provided
by its members. It is now known formally as ACIL: The Association of Independent Scientific,
Engineering and Testing Firms. *ACIL Newsletter* (June 1994), p. 1.
    19.   Personal communication with Jack Moody, Chief, Service Census Branch, U.S. Bureau
of the Census, June 6, 1994.
    20.   U.S. Bureau of the Census, Current Business Reports BS/92, *Service Annual Survey:
1992*, 18.
    21.   ISO, *Compendium of Conformity Assessment Documents*, 154.
    22.   Breitenberg, ed., *Directory of Federal Government Laboratory Accreditation/Designa-
tion Programs*, 52-53.
    23.   ISO, *Compendium of Conformity Assessment Documents*, 155.
    24.   ANSI, *American National Standards for Certification Z34.1*.
    25.   Breitenberg, *The ABC's of Certification Activities in the United States*, 4-5, 12-13.
    26.   For a discussion of types of product certification programs and factors influencing choice
of methods to be used, see ISO, *Certification and Related Activities*, 21-50.
    27.   Breitenberg, *The ABC's of Certification Activities in the United States*, 7.
    28.   ISO, *Certification and Related Activities*, 21-36.
    29.   American Council of Independent Laboratories, Inc., *Directory (22nd ed., 1992-1993)*,
1992.
    30.   Underwriters Laboratories, *An Overview of Underwriters Laboratories: Testing for Pub-
lic Safety*, brochure, 1993.
    31.   Breitenberg, ed., *Directory of U.S. Private Sector Certification Programs*, 75.
    32.   Nina I. McClelland, David A. Gregorka, and Betsy D. Carlton, "The Drinking Water
Additives Program," *Environmental Science & Technology* 23, no. 1 (1989), 14-18; and NSF International, "NSF International," brochure, n.d.
    33.   Breitenberg, *The ABC's of Certification Activities in the United States*, 6.
    34.   Breitenberg, *The ABC's of Certification Activities in the United States*, 7.
    35.   Breitenberg, *The ABC's of Certification Activities in the United States*, 7.
    36.   See Breitenberg, ed., *Directory of Private-Sector Certification Programs*, 46, 167; and
Cheit, *Setting Safety Standards*, 22, 27-28.

37. Breitenberg, *Index of Products Regulated by Each State*.

38. NIST, "Weights and Measures," in *Guide to NIST*, November 1993, available through Internet at gopher://gopher-server.nist.gov.

39. Breitenberg, *The ABC's of Certification Activities in the United States*, 8.

40. Breitenberg, *The ABC's of Certification Activities in the United States*, 8.

41. Cheit, Setting Safety Standards, 14; and Charles Hyer, Publisher, *TMO Update*, personal communication, June 22, 1994.

42.. For the most recent available directory, see Breitenberg, *Directory of Federal Government Certification Programs*.

43. U.S. Department of Defense, pamphlet, *Qualified Manufacturers List (QML): Capturing Commercial Technology for Microelectronics*.

44. Information provided by staff of Office of Standards Services, NIST, May 1994.

45. Deloitte & Touche Management Consulting, *ISO 9000 Survey*.

46. *Quality Systems Update: Special Supplement* 4, no. 5 (May 1994): 4.

47. See Breitenberg, *More Questions and Answers about the ISO 9000 Standard Series and Related Issues*.

48. See, for example, Reimann and Hertz, *The Malcolm Baldrige National Quality Award and ISO 9000 Registration*, 42-53.

49. Michael Barrier and Amy Zuckerman, *Quality Standards the World Agrees On*, 71-73.

50. Because the ISO 9000 standards are relevant only to the manufacturing process and not the product itself, a poor product design could be produced with a high degree of consistency by a high-quality, ISO 9000-compliant manufacturing plant. A ridiculous example that serves to illustrate the concept is a hypothetical, ISO 9000-registered manufacturing plant producing "high-quality" concrete life preservers. For further discussion, see Barrier and Zuckerman, *Quality Standards the World Agrees On*.

51. Unlike ISO 9000, the Malcolm Baldrige National Quality Award (MBNQA) criteria include measures of customer satisfaction. Although very few firms can win the MBNQA in a given year, widespread dissemination of information about the award criteria for firms to follow voluntarily can be an effective means of promoting quality in U.S. industry. See Reimann and Hertz, *The Malcolm Baldrige National Quality Award and ISO 9000 Registration*.

52. ISO, *Certification and Related Activities*, 65-75.

53. For a discussion of the role of quality system registration as part of a comprehensive conformity assessment scheme, see ISO, *Certification and Related Activities*, 24-26, 77-80.

54. Breitenberg, *Q&A on ISO 9000*; and Performance Review Institute, *National Aerospace and Defense Contractors Accreditation Program*, brochure, n.d.

55. Greg Saunders, et al. Working Group on Military Specifications and Standards, *Road Map for Milspec Reform*, 32-33.

56. This standard overlaps closely with ISO 9000 and other quality management system standards. The lack of a mechanism for the Nuclear Regulatory Commission to accept ISO 9000 registration as a substitute for 10 C.F.R. 50, Appendix B, forces parts suppliers to the nuclear industry to undergo duplicate registration procedures. As a result, the supplier base in the industry has been reduced in size and competitiveness. Electric Power Research Institute, "Comparison of ISO 9000 Requirements to Those of 10 C.F.R. 50 Appendix B: Task Sheet", April 8, 1992; and data provided by Lonnie Dunn, Procurement Quality Assurance Office, Illinois Power Company, 1994.

57. "Big Three Standards to be Rolled Out This Month," *Quality Systems Update*, CEEM Information Services newsletter July 1994): 1.

58. Data provided by Lawrence L. Wills, IBM Director of Standards, October, 1993.

59. The survey was sent to all 1,679 U.S. and Canadian firms registered as of the date of the survey. 37 percent of the survey population responded, representing a statistically valid sample of firms in all major industry sectors and a wide range of firm sizes. Foster Finley and Don Swann, *ISO 9000 Survey Results*, Deloitte & Touche Management Consulting, January 1994.

60. The survey found that companies with comprehensive quality management systems already in place in advance of seeking ISO 9000 registration were no more or less likely than other firms to anticipate recovering the expense of registration through cost savings. Foster Finley, Deloitte & Touche Management Consulting, *ISO 9000: Investment or Expense?*, 1993.

61. "GSA Clarifies Stance on ISO 9000 Registration", *Quality Systems Update* 4, no. 5 (May 1994), 9; and "DoD, NASA Officials: Thumbs Up on ISO 9000," *Quality Systems Update* 4, no. 2 (February 1994), 1.

62. Joe Cascio, presentation at 1994 ANSI Annual Public Conference.

63. See Locke, *Conformity Assessment—At What Level?*

64. Locke, *Conformity Assessment—At What Level?*, 5-6.

65. Performance Review Institute, *National Aerospace and Defense Contractors Accreditation Program*; and Charles Hyer, "Performance Review Institute," *TMO Update*, May 20, 1994, 6.

66. Breitenberg, *Laboratory Accreditation in the United States* 5-8; and Charles Hyer ed., *Directory of Professional/Trade Organization Laboratory Accreditation/Designation Programs*, 3-5.

67. See NIST, *National Voluntary Laboratory Accreditation Program: Procedures and General Requirements.*

68. NIST, *National Voluntary Laboratory Accreditation Program: Fee Schedule and Worksheets.*

69. A2LA, *1993 Annual Report.*

70. A2LA, "Application for Accreditation", 1994.

71. See Locke, *Conformity Assessment—At What Level?*, 1-4.

72. Breitenberg, *Laboratory Accreditation in the United States*, 15.

73. U.S. General Accounting Office, *Laboratory Accreditation: Requirements Vary Throughout the Federal Government*, 26-28.

74. Data supplied by Charles Hyer, Executive Vice President, The Marley Organization, August 1994.

75. Kim Phillipi, Entela, Inc. to Joseph O'Neil, ACIL, memorandum, 19 August 1994, *Accreditation Costs* (Grand Rapids, Mich., 1994); David Krashes, MMR, Inc. to Joseph O'Neil, ACIL, West Boylston, Mass., 11 August 1994, letter; and Walter Poggi, Retlif Testing Laboratories to Joseph O'Neil, ACIL, memorandum, n.d., *Retlif Testing Laboratories' Yearly Accreditation Costs.*

76. See U.S. Environmental Protection Agency, Jeanne Hankins, ed., *Final Report of the Committee on National Accreditation of Environmental Laboratories.*

77. For details of results and methodology, see U.S. Environmental Protection Agency, *Final Report of the Committee on National Accreditation of Environmental Laboratories.*

78. NIST, "Establishment of the National Voluntary Conformity Assessment System Evaluation Program." In *Federal Register*, 19129-19133.

79. See NIST, *Establishment of the National Voluntary Conformity Assessment System Evaluation Program.*

80. Internationally accepted standards for conformity assessment services, such as testing, certification, and accreditation, are outlined in a series of guides published by the ISO Council Committee on Conformity Assessment (CASCO). See ISO, *Information on CASCO.*

81. NIST, "Establishment of the National Voluntary Conformity Assessment System Evaluation Program," 19131-19132.

82. *Fastener Quality Act*, 101st Congress, 2nd session, H.R. 3000.

4

International Trade

O ver the past decade, rising exports have become a principal source of U.S. economic growth. The expansion of global trade in the postwar era has promoted higher standards of living worldwide. Today, many nations around the globe have made great progress in moving toward market systems based on the principles of open trade and investment. For U.S. exporters seeking to take advantage of expanding opportunities, the relationship among standards, conformity assessment, and global trade is increasingly important. Since international trade constitutes a growing share of highly specialized production and services in the U.S. economy, for example, barriers to trade reflected in discriminatory standards and conformity assessment systems threaten to retard U.S. economic progress.

Considerable progress has been made since the Second World War in lowering international trade barriers, particularly those associated with tariffs. As tariff barriers have decreased, however, the relative significance of non-tariff barriers to trade, including those related to standards, has increased. In 1994, a major multilateral trade agreement was concluded in the Uruguay Round of the General Agreement on Tariffs and Trade (GATT). The Uruguay Round made significant progress in addressing the rise of non-tariff trade barriers. Strengthening of GATT provisions on standards and conformity assessment-related barriers to trade, combined with the establishment of new enforcement mechanisms through the World Trade Organization (WTO), indicate the potential for significant progress in facilitating U.S. exports and future economic growth.

Realizing the full benefits of these opportunities, however, will require creative, forward looking, and aggressive U.S. trade policies. This will involve work to support (1) implementation of the multilateral standards agreements under the WTO; (2) innovative new efforts to link standards and U.S. export promotion services; (3) policies and programs to ensure systems surveillance in areas of growing significance to trade, especially environmental management systems standards and national conformity assessment regimes; and (4) possible aggressive, unilateral use of U.S. law to counteract unfair foreign trade practices in standards.

One of the most significant advances in the Uruguay Round, for example, was the expansion in the number of countries brought under disciplines on standards first negotiated in the Tokyo Round of the GATT. Many U.S. trading partners, especially the developing nations of Asia and Latin America, will require assistance in constructing modern standards and conformity assessment regimes. By providing this assistance, the United States has an opportunity to support successful implementation of the GATT, as well as U.S. global export expansion and economic progress.

In the context of these developments, issues concerning standards and conformity assessment have moved to a central position in future U.S. foreign economic policy. This chapter outlines the link among product and process standards, global trade, and U.S. economic interests. As discussed in this chapter, our national trade policy objectives are served through the removal of technical trade barriers in key export markets; participation in international, mutual recognition of conformity assessment systems; and expanded efforts at export promotion through cooperation and assistance to standards bodies in existing and emerging U.S. export markets.

STANDARDS, TRADE, AND
U.S. ECONOMIC PROGRESS

Trade and the expansion of global exports are directly linked to U.S. economic vitality and future standards of living. Exports provide for domestic economic growth, increased labor productivity, and creation of jobs in the manufacturing and service sectors that pay wages well above the national average.[1] Future U.S. economic success, as a result of these factors, centers to an increasing extent on removing barriers to international trade, as well as creating innovative export promotion programs to help expand markets for U.S. goods and services overseas.

Standards and the Economic Benefits of Trade Expansion

There are many indicators of the importance of exports to the domestic economy. U.S. exports have grown at a rapid rate as a percentage of Gross Domestic Product (GDP) over the past decade. Goods and services exports rose from 7.5 percent of GDP in 1986 to approximately 13 percent in 1993.[2] Merchandise exports alone rose from 5 percent of GDP in 1984 to 8 percent in 1993. Total U.S. exports have more than doubled over the period 1985-1993, from $218 billion to $464 billion.

In particular, merchandise exports of advanced technology products have risen sharply. Many of these exports are directly affected by international standards, as well as technical regulations of governments overseas. Strong export growth in advanced technology goods resulted in a U.S. trade surplus in these accounts at $25.8 billion.[3] This surplus has offset trade deficits in non-advanced technology products each year since 1982. The estimated non-advanced technology U.S. trade deficit was $141.6 billion in 1993, with a total merchandise trade deficit estimated at $115.8 billion. Moreover, advanced technology product exports rose as a share of total merchandise exports to 23 percent in 1993.

The rapid expansion of exports has played an important part in U.S. employment growth over the past decade. As of 1990, there were 7.2 million U.S. workers employed in export-related jobs. This represented 20 percent of the total 10.4 million job increase in U.S. employment over the period 1986-1990. Moreover, jobs linked to exports paid wages on average 17 percent higher than the national average for all U.S. workers in 1990 ($11.69 versus $10.02, respectively).[4] It is clear that to the extent the U.S. continues to pursue a trade policy focused on the opening of global markets and trade expansion, such a policy will provide greater employment opportunities in high-paying jobs.

The benefits of open markets and specialization are important to multinational firms with operations across the globe. They also benefit small and medium-sized firms, whether they export directly or supply components that other firms incorporate into exported products. As outlined in Chapter 1, standards help foster economies of scale in the production process. Economies of scale through open trade allow wide consumer choice and increased purchasing power and consumer welfare.

Access to foreign markets also maximizes benefits available through the globalization of production, including access to diversified sources of technology, manufacturing advances, and information on best practices in marketing, sales, and service. In addition, through increased market size on a global scale, firms are able to spread the costs of investment in research and development across larger numbers of sales. In sum, open trade provides the platform through which firms and nations can leverage the benefits of competitive forces in support of long-term economic advance, consumer welfare, and productivity growth.

Standards and conformity assessment are closely linked to these benefits of international trade. Standards development systems and the infrastructure necessary to ensure conformity to standards—including testing, certification, and laboratory accreditation—are an important part of modern industrial processes, as discussed in previous chapters. In general, the benefits of standards observed in the domestic economic context increase in proportion to their application on an increasingly larger, international scale.

Efficient international standards regimes accomplish several important goals. These include facilitating the diffusion of innovative technologies and production techniques, as well as supporting global economies of scale.[5] When different countries or regions have different technical standards for essentially the same product, manufacturers selling into multiple markets are forced to produce multiple versions of the same product. For example, automobile production lines must be switched between right-hand and left-hand drive cars for the United Kingdom and continental Europe. Consumer electronic devices and household appliances must be adapted for different power supplies in the United States and Europe—110 and 220 volts, respectively.

By fragmenting the prospective markets for products that could otherwise be produced and marketed on a global scale, the lack of internationally harmonized standards reduces the economic advantages of free-flowing international markets. Harmonization entails the revision or interpretation of different standards in such a way as to render them equivalent. International harmonization of standards enables manufacturers to produce more efficiently for a larger, combined market.

In addition to promoting economies of scale to facilitate multiple export markets, common standards and conformity assessment procedures benefit manufacturers in other ways. Standards convey information to customers about products and services in a technically precise, consistent manner, as outlined in Chapter 1. This is an important benefit when manufacturer and customer are separated across linguistic, cultural, and geographic distances. Well-organized, open, and transparent standards systems also promote compatibility of key components in national infrastructure, such as telecommunications and computer networks.[6] Finally, standards and technical regulations can operate to support public welfare by promoting health, safety, and environmental goals. These affect not only the quality of domestic industrial production, but also the operation of international markets and U.S. export success within those markets.

Conformity assessment procedures are also directly linked to the efficient functioning of international markets. *Even when standards in different countries have been harmonized, the free flow of trade is inhibited if products are subjected to redundant testing and certification requirements in multiple export markets.* When conformity assessment procedures performed within the United States are not accepted as valid by regulators or purchasers in foreign markets, U.S. exporters are forced to ship products abroad for costly, wasteful retesting. They also may have to support the costs associated with bringing foreign inspectors to visit and inspect U.S. manufacturing facilities. When nations, states, or local governments here in the United States refuse to accept competent and scientifically sound testing and certification performed abroad without reasonable justification, the costs of imported goods are raised in a discriminatory manner. This is true whether the underlying standards for a product are harmonized between an exporter's home and final destination.

Cost of Protection: Non-Tariff Barriers to Trade

Trade protection that restricts competition or restrains circulation of products in international markets reduces global efficiency and slows economic advance. A number of policy tools are used by nations to shelter firms from international competition.[7] These include high tariff rates; voluntary export restraints; production subsidies; import quotas; and a wide range of non-tariff barriers, including those related to standards and conformity assessment mechanisms. Standards that discriminate against imports and nontransparent or discriminatory requirements for showing conformity to standards can create significant non-tariff trade barriers.

The economic harm caused by trade discrimination and protection of domestic markets is well documented.[8] Most empirical research and data on the costs of instruments to either block imports or subsidize exports have focused on indirect measures of trade protection. Their effect on increased consumer prices and reduction of global wealth is clear. A recent study of the cost of protection in textiles, apparel, autos, and steel, for example, found that removal of all quantitative restrictions, such as quotas, on imports into the United States would result in a 0.5 percent increase in national income. This increase would be worth $25 billion to $29 billion dollars. The same study found that the reduction in welfare caused by U.S. non-tariff barriers in these sectors is equivalent to that which a 49 percent tariff would produce.[9]

Great progress has been made in reducing worldwide tariffs since the Second World War. Prior to 1947 and the establishment of the GATT, for example, average weighted tariffs on goods in the industrialized nations stood at 35 percent. A tax of $35 per $100 of products traded before 1947, therefore, was paid by final goods producers on imported components and by consumers of finished products. After the Tokyo Round of GATT trade negotiations ended in 1979, average tariffs in the major trading markets of the United States, Japan, and Europe were lowered to about 3.8 percent. The Uruguay Round of negotiations completed in 1994 cut tariffs in major industrial markets to zero in many sectors. These include construction, agricultural, and medical equipment; pharmaceuticals; paper; toys; and furniture, among others.[10] Tariff cuts ranged from 50 to 100 percent on semiconductors and computer components. The agreement also committed many emerging, newly industrializing markets to cut tariffs sharply in these sectors.

Although global tariffs have been reduced, there has been a rise in the use of other mechanisms to deny access of goods to national markets. Whereas the extent and costs of traditional forms of trade protection are well documented, less attention has been devoted to analyzing or measuring the effects of non-tariff barriers to trade.[11] This is particularly true of analysis on the trade effects of discriminatory standards and conformity assessment procedures.[12] It is clear, however, that non-tariff barriers raise costs of production in a manner similar to tariffs, by increasing the price of imported materials and components and reducing a manufacturer's ability to profit from economies of scale. At the same time, they block more competitive and efficient producers from reaping the benefits of superior products through expansion of sales in export markets. In many cases, especially in the capital goods sector, the world's most efficient and competitive producers are U.S. firms.

Although the definitions of non-tariff barriers varies among academic studies, the consensus of several studies is that they are spreading worldwide.[13] A study by the World Bank in 1987, for example, found that the share of imports from industrialized nations subject to "hard-core" non-tariff barriers in the Organization for Economic Cooperation and Development (OECD) nations rose from 13 to 16 percent from 1981 to 1986. These hard-core barriers included import prohibitions, quantitative restrictions, voluntary export restraints, variable levies, restrictions on textiles and apparel, and nonautomatic licensing.[14] Since 1970, the United States, Europe, and Japan have been the most active in implementing these restrictions. To the extent that increases in non-tariff barriers offset the gains achieved in the past few decades from lowering tariffs, this trend is a cause for serious concern.

To the detriment of sound public policymaking, particularly as it relates to future trade negotiations, there have been no comprehensive analyses of standards and conformity assessment systems as non-tariff barriers to trade. There is a significant and growing need for academic and policy-oriented studies and research in this area. Within the context of the work related to the general subject of non-tariff barriers to trade, however, there is evidence to indicate that significant barriers to global trade are embedded in existing standards and will continue to grow in complexity.

This conclusion is based, in part, on observations such as the following: (1) standards that differ from international norms are employed as a means to protect domestic producers; (2) restrictive standards are written to match the design features of domestic products, rather than essential performance criteria; (3) there remains unequal access to testing and certification systems between domestic producers and exporters in most nations; (4) there continues to be a failure to accept test results and certifications performed by competent foreign organizations in multiple markets; and (5) there is a significant lack of transparency in the systems for developing technical regulations and assessing conformity in most countries. Moreover, observations from U.S. government and industry sources indicate that domestic firms continue to confront problems associated with systems in overseas markets, as discussed in the following section.[15]

Barriers to Trade in Key U.S. Export Markets

The National Trade Estimate (NTE) Report on Foreign Trade Barriers, produced annually by the Office of the U.S. Trade Representative (USTR), outlines foreign use of discriminatory standards, testing, and certification requirements as barriers to U.S. exports.[16] In a number of foreign markets, U.S. goods are subject to more stringent standards and testing requirements than domestic products (for selected examples, see Table 4-1). Many of the barriers to U.S. products outlined in the NTE reports affect sectors in which U.S. industry enjoys substantial comparative advantages over foreign competitors. This is true, for example, in the capital goods and high-technology sectors, including transportation equipment, electrical machinery, medical devices, biotechnology, and pharmaceutical products, among others. In the absence of trade barriers, these are sectors in which U.S. firms have strong potential to gain export market share.

One example of foreign technical regulations as unfair barriers to U.S. exports is the case of the European-wide ban on use of livestock growth hormones.[17] In 1989, the European Union (EU) banned the import of meats and meat products, except pet food, produced with the aid of natural or synthetic growth hormones. The U.S. considers this a trade violation. An increasing number of U.S. producers employ these hormones. There is no internationally recognized scientific evidence to support EU regulation. The ban eliminated most U.S. red meat and meat product exports, causing $97 million per year of economic harm. In retaliation, the United States imposed an equal value of tariffs on EU agricultural products and brought the dispute for resolution to the GATT. As of September 1994, the EU has successfully blocked resolution of the GATT case. However, new dispute procedures instituted in the Uruguay Round agreements, discussed later in this chapter, will lead to more rapid resolution of cases such as this.

Data on the total volume of global trade subject to conformity assessment regulations are limited. Increased attention by U.S. government agencies to data gathering and analysis would be of significant benefit in formulating sound U.S. trade policies, as well as supporting U.S. export promotion programs. As noted previously, conformity assessment systems have the potential to create equal or greater barriers to trade than standards. Worldwide growth in the complexity of mechanisms for approval of regulated products, such as food additives and medical devices, is of particular concern. As discussed in Chapter 3, government agencies and private-sector firms in the United States and abroad are involved in performing redundant testing, certification, quality system registration, and laboratory accreditation. If not monitored and addressed in a systematic manner, these systems will provide a great number of opportunities for nations to employ a variety of extremely complex and nontransparent barriers to imported goods.

There have been only a limited number of recent attempts to estimate the impact of standards and conformity assessment barriers on U.S. trade. The Department of Commerce and Trade Policy Coordinating Committee (TPCC) have, however, completed preliminary work in this area. The project could not independently verify the department's analysis, because supporting data and a detailed methodology employed in the work were unavailable for review. The

TABLE 4-1 — Standards and Certification: Selected Barriers to U.S. Exports

COUNTRY	TRADE BARRIERS
European Union (EU)	In 1989, the EU banned meat and meat products produced from livestock treated with natural or artificial (biotechnology-derived) growth hormones. The EU has stated its recognition that there is no scientific evidence to support the ban; however, it remains in effect. Damage to U.S. exporters is measured at $97 million per year. *STATUS*: Unresolved; U.S. retaliation under Section 301 remains in effect.
	The EU's "Global Approach to Testing and Certification," instituted in 1990, mandates that certification of regulated products be performed by European testing laboratories and certifiers. This system imposes unbalanced costs on non-European manufacturers for obtaining product approvals. (U.S. government accreditation, by contrast, does not discriminate between U.S. and foreign laboratories.) In some sectors, testing can be performed by a U.S. laboratory under subcontract to a European laboratory. Success in U.S.-EU mutual recognition agreement (MRA) negotiations, which began in 1994, would remove or reduce this barrier. *STATUS*: MRA negotiations ongoing.
Japan	Prescriptive design standards under the High Pressure Gas Law favor Japanese producers. Affected sectors include air conditioners, refrigeration equipment, supercomputers, and aircraft support equipment. Performance standards would be less trade restrictive and more flexible in accommodating technological innovations. *STATUS*: Ongoing.
	Barriers to imported wood products were estimated, in 1989, to restrict U.S. exports between $500 million and $2 billion annually. Key among the barriers were restrictive fire and building codes and refusal to accept foreign testing procedures. The National Forest Products Association petitioned the USTR to initiate a Super 301 case against Japan. *STATUS*: Case was resolved in April 1990. Japan agreed to increase reliance on performance-based standards and to accept foreign test data. International technical committees monitor implementation, which has been largely successful.
China	China does not accept U.S. certifications of quality. Procedures for obtaining a required "quality license" are costly and discriminatory, often imposing higher standards for imports than domestic goods. Many regulatory requirements are unknown or unavailable to non-Chinese firms. *STATUS*: In 1991, USTR self-initiated a Section 301 investigation of these and other Chinese trade practices. China agreed to publish notice of regulations and to discuss other issues. Implementation remains unclear.
Indonesia	Acceptance of new pharmaceutical imports can take more than a year. Copied products are often available on the local market before the original is accepted. *STATUS*: Ongoing.
Republic of Korea	Many regulatory standards, such as shelf-life standards for processed foods, differ substantially from international practices without scientific basis. Public notice of rule making is often inadequate. Some standards are applied unequally to imported and domestic products. Medical equipment and processed agricultural products are among imports facing nontransparent or unclear standards. *STATUS*: Ongoing.
Taiwan	Agricultural imports are routinely tested, unlike domestic products. Complex registration procedures exist for approval of imported pharmaceuticals, medical devices, and cosmetics. *STATUS*: Ongoing.

TABLE 4-1 — (continued)

COUNTRY	TRADE BARRIERS
Mexico	Beginning in July 1994, regulated products must be tested and certified by laboratories accredited by the Mexican Director General for Standards (DGN) or by DGN itself. Quality system registrations must be performed by DGN-accredited ISO 9000 registrars. In practice, DGN has accredited no non-Mexican laboratories or registrars and very few Mexican ones. It is not obligated to accredit or recognize foreign organizations, under the North American Free Trade Agreement, until 1998. *STATUS*: Ongoing.

SOURCES: U.S. Trade Representative, Office of the. *1994 National Trade Estimate Report on Foreign Trade Barriers*. Washington, D.C.: U.S. Government Printing Office, 1994.

Bayard, Thomas O. and Kimberly Ann Elliott. *Reciprocity and Retaliation in U.S. Trade Policy*. Washington, D.C.: Institute for International Economics, 1994.

American National Standards Institute. *New Certification Methods Established for Exports to Mexico That Are Subject to Mandatory Standards*. 1994.

Retlif Testing Laboratories. *U.S./EU Trade Negotiations*. 1994.

department indicates, however, that $300 billion of the $465 billion in U.S. merchandise exports in 1993 were affected by foreign technical requirements and standards. A total of $180 billion is reportedly subject to certification to non-U.S. standards in such sectors as automotive, aerospace, computers, telecommunications, pharmaceuticals, and chemicals. An additional $70 billion was subject to quality or environmental management system registration.[18]

These estimates and the analysis supporting the work need to be carefully examined and reviewed. They provide some indication, nevertheless, that the addition of new layers of complexity and cost in international commercial transactions is cause for concern. There is danger that a proliferation of complex, costly, and redundant conformity assessment systems among nations will present serious problems in future international trade. Duplicative or discriminatory requirements threaten to undermine the trade-enhancing benefits of international standards by adding layers of costly and redundant requirements for showing conformity to specifications in multiple export markets.

The request by Congress for this study specifically identified the evolving product approval systems in the European Union as a source of concern for the United States. For example, the expense of meeting EU approval requirements can be a particular obstacle to small and medium-sized U.S. firms. This is true whether the firm exports directly to European markets or acts as a supplier of components to manufacturers of exported goods.[19] With Europe as the largest single destination for U.S. exports, accounting for 25 percent of total U.S. merchandise exported, European barriers to U.S. products have a direct and substantial impact on U.S. trade performance.[20]

According to estimates by the Department of Commerce's International Trade Administration, $66 billion of the $110 billion in U.S. merchandise exports to Europe in 1993—more than half—was subject to some form of EU-required product certification. Approximately $30 billion required government-issued certificates, with pharmaceuticals, automobiles, and engines accounting for the majority. An additional $25 billion required a manufacturer's declaration of conformity (self-certification), while $10 billion was subject to private, third-party certification—primarily in the information technology sector.[21]

The transition to a unified European economic market has included a number of measures to remove barriers to trade in regulated products among European nations. To the extent that market unification enables U.S. exporters to access all nations of the European Union by meeting the import requirements of any single European country, these changes facilitate increased U.S. exports. New EU requirements for product approval, however, have raised serious concern about U.S. access to the European market. Changes in EU procedures for setting standards and verifying product compliance with them, their potential effect on U.S. trade, and the utility of U.S.-EU agreements on conformity assessment in mitigating these effects are discussed in detail in the section on mutual recognition negotiations.

A forceful effort at streamlining international systems in standards, certification, and quality regulations at the regional and multilateral levels is an appropriate priority for U.S. trade and standards policymakers. The goal of these efforts is clear. They should provide a mechanism for manufacturers servicing global markets to obtain testing, certification, and registration of quality systems one time, and in one market, to have products accepted globally. The goal of assured credibility in these systems is also essential. Reaching this goal should be a multilateral priority, however, work to advance this principle can be undertaken unilaterally and at the regional level, as outlined later in sections on the use of U.S. trade law to lower barriers and regional dialogue on mutual recognition agreements (MRAs) in the Asia Pacific Economic Cooperation Council (APEC) forum. As the next section outlines, international mechanisms also exist to reduce standards-related trade barriers. These mechanisms are part of the GATT and new World Trade Organization created through the Uruguay Round of multilateral negotiations.

MULTILATERAL TRADING SYSTEM: THE URUGUAY ROUND

The Uruguay Round of multilateral trade negotiations under the General Agreement on Tariffs and Trade concluded in 1994 with the signing of a world trade agreement.[22] Significant progress was made in advancing the goal of reducing barriers to trade, including both tariffs and non-tariff barriers. Members accepted a revised Agreement on Technical Barriers to Trade (TBT). The TBT agreement, or Standards Code, was first incorporated in the previous Tokyo Round of the GATT.[23] A new Agreement on Sanitary and Phytosanitary Standards (SPS) was also concluded, with special implications for global agricultural trade.[24] President Clinton signed the U.S. law to implement these agreements on December 8, 1994. It is expected that the GATT obligations will enter into force on January 1, 1995, for all member nations.

Table 4-2 describes major advances in coverage related to standards and conformity assessment in the Uruguay Round Agreements.[25] The TBT agreement covers "product characteristics or their related processes and production methods," as reflected in mandatory technical regulations of national governments. The agreement attempts to reduce barriers to trade reflected in the preparation, adoption, or application of standards in a discriminatory manner. It also addresses the prevention of new barriers, particularly as they might arise in divergent conformity assurance systems. The TBT agreement also has important implications for standards set by subnational and regional governments (such as the EU) and private-sector bodies. This section outlines the key elements of the agreement to reveal progress made in the Uruguay Round in relationship to the existing Standards Code negotiated in the Tokyo Round, as well as areas of uncertainty in its implementation and impact on trade.[26]

Membership and Expansion of Scope

The most important progress made in the Uruguay Round TBT and SPS agreements is related directly to the expanded scope and coverage of international disciplines on technical regulations as they affect trade. This includes both the increase in the number of countries bound by the obligations in the new agreement and the extension of TBT rules to cover new areas of standards and conformity assessment systems.

The expansion in coverage of the Uruguay Round TBT code to include all members of the newly established World Trade Organization is an important move toward strengthened international discipline (see Table 4-3). As of November 1993, there were 46 signatories to the Tokyo Round TBT code. Most of these members are the industrialized nations of the European Union, with the United States and selected Asian and Latin American countries represented. The signatories to the Uruguay Round Agreement and the new TBT agreement include 68 additional nations. Many of these are among the most rapidly developing nations of Asia and Latin America. Based on 1991 data, new signatories of the Uruguay Round TBT agreement represent an expansion of approximately $182 billion in global imports subject to international discipline. This is a 17.5 percent increase over imports covered under the Tokyo Round Standards Code.[27]

The extension of rules and procedures on standards and conformity assessment is an important part of strengthening the multilateral trading system. The TBT code helps support progress toward global market liberalization worldwide.

TABLE 4-2 — Standards, Conformity Assessment, and the GATT: Key Provisions of the Tokyo and Uruguay Rounds

KEY PROVISIONS	TOKYO ROUND [completed 1979]	URUGUAY ROUND [completed 1994]
World Trade Organization (WTO)	Prior to the Uruguay Round, enforcement of GATT disciplines was through the GATT Secretariat. Not all signatories to the GATT were parties to the Standards Code. This led to a "free-rider" problem. Dispute settlement process lacked "teeth."	Created a new body, the WTO, as the administering body of the international trading system. *Membership is available only to countries that are signatories to the GATT* and agree to adhere to all of the Uruguay Round agreements.
Technical Barriers to Trade (TBT) Agreement — Standards Code	The GATT negotiations produced a series of new international agreements relating to non-tariff measures, including the TBT Agreement (the Standards Code). The Standards Code's purpose was to ensure that technical regulations and standards would not be prepared, adopted, or applied with a view to creating obstacles to international trade. Imposes no obligation to lower standards for public health, safety, or environmental protection. Obligation is to treat foreign and domestic products without discrimination, and to use international standards where possible and appropriate.	The requirement of transparent and nondiscriminatory procedures for issuing product approval *was expanded to cover the range of conformity assessment procedures*, including testing, certification, accreditation, and quality system registration. Encourages mutual recognition of conformity assessment procedures between countries. Expands coverage to nongovernmental and regional standards development. Same as Tokyo Round. In addition, states explicitly: "Members shall give positive consideration to accepting as equivalent technical regulations of other Members, even if these regulations differ from their own, provided they are satisfied that these regulations adequately fulfill the objectives of their own regulations."
Sanitary and Phytosanitary Measures	Not a separate agreement.	A stand-alone agreement. For the first time, *establishes multilaterally recognized rules and disciplines for the development and application of measures taken to protect human, animal, or plant life* or health in the areas of food safety and agriculture.

	Standards Code	TBT and SPS Agreements
Applicability to GATT members	Standards Code (Technical Barriers to Trade Agreement) was a stand-alone agreement with its own institutional provisions (e.g., dispute settlement procedures). GATT members were not obligated by the code unless they specifically accepted it. As of 1993, 46 GATT members had signed the code.	*The TBT and SPS agreements are binding on all members of the WTO.* They impose obligations on the members' central governments, local governments, nongovernmental bodies, and international and regional standardization and conformity assessment bodies. They call for national governments to take "reasonable measures" to ensure compliance with the disciplines in the agreement by subnational and nongovernmental bodies. As of 1994, more than 120 nations had signed the WTO agreement.
Standards-related activities: definition and coverage		
Technical regulations	Technical specifications with which *compliance is mandatory.* Signatories to the Standards Code are required to notify other members of proposed new or revised technical regulations.	Explicitly states that technical regulations *should not be maintained if the circumstances or objectives giving rise to their adoption no longer exist* or if the changed circumstances or objectives could be addressed in a less trade-restrictive manner.
Standards	Technical specification approved by a recognized standardizing body for repeated or continuous application, with which *compliance is not mandatory.*	*Expanded to explicitly include standards for processes as well as products.* A standard is defined as a document approved by a recognized body that provides, for common and repeated use, rules, guidelines, or characteristics for products or related processes and production methods, with which compliance in not mandatory. It may also include or deal exclusively with terminology, symbols, packaging, marking, or labeling requirements as they apply to a product, process, or production method.

TABLE 4-2 — (continued)

KEY PROVISIONS	TOKYO ROUND [completed 1979]	URUGUAY ROUND [completed 1994]
Standards-related activities: definition and coverage (continued)		
Conformity assessment	Only covered certification and testing. Parties were not to discriminate against imports in cases where positive assurance was required that products would conform with technical regulations or standards. Parties agreed, whenever possible, to accept test results and certifications issued by relevant bodies in the territories of other parties.	The requirement of transparent and nondiscriminatory procedures for issuing product approval *was expanded beyond testing and certification to cover the range of conformity assessment procedures, including laboratory accreditation and quality system registration.* Conformity assessment is defined as any procedure used, directly or indirectly, to determine that relevant requirements in technical regulations or standards are fulfilled.
Authority and dispute resolution	Procedures were specific to the Technical Barriers to Trade Agreement. These were not identical to procedures under other GATT agreements. There was no legal obligation to carry out dispute resolution panel findings. Ample opportunities existed to delay the resolution process.	A single panel addresses all issues under any of the GATT/WTO agreements. The procedure was consolidated in the Understanding on Rules and Procedures Governing the Settlement of Disputes, and applies to the entire GATT and its subagreements, including those related to standards. It improves the existing system by *providing strict time limits for each step in the dispute settlement process.* The automatic nature of the new procedures *improves enforcement of the substantive provisions.* In certain cases, cross-retaliation measures are authorized.

SOURCES: U.S. Department of Commerce, International Trade Administration fact sheets available on the Uruguay Round Hotline.

Jackson, John H. and Davey J. William. *Legal Problems of International Economic Relations.* Second edition. St. Paul, Minnesota: West Publishing Company, 1986.

TABLE 4-3 — GATT Members Subject to Standards Provisions

Tokyo Round: Standards Code Signatories
(as of November 1993)

Argentina	France	Luxembourg	Singapore
Australia	Germany	Malaysia	Slovak Republic
Austria	Greece	Mexico	Spain
Belgium	Hong Kong	Morocco	Sweden
Brazil	Hungary	Netherlands	Switzerland
Canada	India	New Zealand	Thailand
Chile	Indonesia	Norway	Tunisia
Czech Republic	Ireland	Pakistan	United Kingdom
Denmark	Israel	Philippines	United States
Egypt	Italy	Portugal	Yugoslavia
European Union	Japan	Romania	
Finland	Korea, Republic of	Rwanda	

Uruguay Round Signatories
(as of April 1994)

All of the above,[a] plus the following:

Angola	Costa Rica	Liechtenstein	Qatar
Algeria	Cote d'Ivoire	Macau	Saint Lucia
Antigua and Barbuda	Cuba	Madagascar	Senegal
Bahrain	Cyprus	Malawi	South Africa
Bangladesh	Dominican Rep.	Mali	Sri Lanka
Barbados	El Salvador	Malta	Suriname
Belize	Fiji	Mauritania	Tanzania
Benin	Gabon	Mauritius	Trinidad and
Bolivia	Ghana	Mozambique	Tobago
Botswana	Guatemala	Myanmar	Turkey
Brunei Darussalam	Guinea-Bissau	Namibia	Uganda
Burundi	Guyana	Nicaragua	United Arab
Cameroon	Honduras	Niger	Emirates
Central African Republic	Iceland	Nigeria	Uruguay
China	Jamaica	Paraguay	Venezuela
Colombia	Kenya	Peru	Zaire
Congo	Kuwait	Poland	Zambia

[a] Except Rwanda and Yugoslavia

It furthers the process of binding the developing nations with their industrialized trading partners in an area of increasing importance to the trade system. This is perhaps particularly important in the APEC region since the economies of East Asia, excluding Japan, are expected to grow 6.2 percent per year from 1993 to 2000 and to constitute 27.9 percent of global GDP by the year 2003.[28]

In addition, the new TBT agreement subjects processes and production methods to the same rules as those applied under the Tokyo Round code to manufactured goods. This expansion of multilateral rules will serve to reduce the likelihood that unjustified measures to block imports through technical regulations will continue unchallenged. There have been several high-profile trade disputes in this area over the past five years. Most notably, these have involved U.S.-EU trade in agricultural products and the EU's Third Country Meat Directive and Beef Hormones Directive, outlined above. At a minimum, a transparent, multilateral framework through the TBT agreement now exists to address these procedures.

Coverage of Conformity Assessment

Technical regulations for oversight of testing and certification represent one of the fastest-growing segments of the industrialized nations' standards systems. Although most of the attention in trade policy discussions has addressed harmonization of differing national standards, expansion of national regulations on *conformity to standards* is where costs to manufacturers and exporters are likely to grow in the future. In the United States, as discussed in Chapter 3, the independent testing industry represented approximately $10 billion in revenue in 1993, with average annual growth rates of 13.5 percent from 1985 to 1992.[29] The growth in third-party testing, moreover, stimulates the growth of added layers of complexity, reflected in public and private programs to accredit testing laboratories and government mechanisms for oversight of both laboratories and accreditors.

In the developing nations, continued industrialization and competition for access to global export markets will likely contribute to expansion in testing and accreditation systems over the next decade. Manufacturers in developed and industrializing nations will be subject to the full range of conformity assessment steps in export markets. Third-party firms that perform these tasks are already displacing the use of manufacturers' self-declarations of conformity in a number of sectors, most notably capital goods.[30] Protocols at the international level to bring clarity and foster acceptance of these systems across national borders will be increasingly important to trade.

The Tokyo Round Standards Code applied the principles of national treatment and nondiscrimination only to product testing and certification programs. These principles require that the treatment of imports by a GATT member country be no less favorable than the treatment of domestic products or imports from other GATT members. Articles 5 through 9 of the Uruguay Round TBT agreement extend the basic obligation of national treatment and nondiscrimination to laboratory accreditation, recognition, and quality system registration programs, such as ISO 9000 registration requirements. Extension of the coverage in the Uruguay Round to all forms of conformity assessment, therefore, sets in place a framework that can serve to protect against their future use as barriers to trade.

Although progress was made in extending national treatment and nondiscrimination to a wider range of conformity assessment, the TBT agreement provides only a limited basis to encourage acceptance of the results of tests or laboratory accreditation across national borders. Article 6 of the TBT code exhorts signatories to move toward harmonization of conformity assessment through mutual recognition of one another's procedures. Mutual recognition agreements have the potential to offer real benefits in reducing costs, inefficiencies, and barriers in international trade. Recent developments at the regional level of MRAs in conformity assessment, including early dialogue within the APEC forum, are outlined later in this chapter.

Extension of Coverage to Nongovernmental Organizations

Another important way in which the TBT agreement represents progress involves the extension of rules to private standards organizations, such as the American National Standards Institute (ANSI) and European regional standards developers. Article 3 of the agreement calls for "reasonable measures" by members of the WTO to ensure compliance by these bodies with principles of national treatment, nondiscrimination, and notification of standards preparation in advance of promulgation. The article further states that "members shall formulate and implement positive measures and mechanisms in support of observance of the provisions of Article 2 by other than central government bodies." Central government will now be responsible for good-faith implementation of the agreement and application of its principles at any level of government or within any private-sector body involved in the standards system. The TBT code of the Tokyo Round bound only central governments and, less rigidly, subnational (state and provincial) governments to these obligations.

The new "Code of Good Practice for the Preparation, Adoption and Application of Standards" contained in Annex 3 of the TBT agreement provides a foundation for extending rules to private standards bodies. The code outlines general principles for development and applications of standards by nongovernmental organizations. These include national treatment of products from a foreign country no less favorable than that accorded to domestic products or imports from any other country (national treatment and nondiscrimination); publication and dissemination of work in progress; institution of a 60-day open comment period prior to adoption of standards; and refraining from applying standards that could serve as barriers to international trade.

Adoption of the Code of Good Practice is voluntary and lacks an enforcement mechanism. It does, however, outline for the first time in a multilateral agreement a common mode of operation for private standards bodies consistent with open trade. Wide acceptance of the code among WTO members also has the potential to foster communication between national standards organizations. This could open new channels for early dialogue and informal dispute resolution, helping to support the more formal, multilateral mechanisms available through the WTO. Finally, nongovernmental standards bodies will have no standing at the WTO or access to dispute settlement on their own authority. These groups will for the first time, however, be able to publicly hold other organizations accountable to standards of conduct and fair access, as a function of the new WTO Standards Agreement.

Dispute Settlement

A crucial point of progress in the Uruguay Round Agreement, lacking in the disciplines of the Tokyo Round Standards Code, is a binding framework for the adjudication of disputes. Trade actions involving technical barriers will be considered as part of an integrated dispute resolution system under the WTO. Violations or noncompliance with TBT provisions found by dispute settlement panels of the WTO will require action to curtail trade-distorting behavior. If not resolved, members will have the right to impose retaliatory tariffs against nations found in violation of the TBT agreement. These provisions should transform the manner in which TBT principles are viewed by governments. It is reasonable to assume that, in the future, some consideration will be undertaken of how national policy changes in standards and conformity assessment rules will affect trade. Possible violations in these areas will involve internationally public consequences.[31]

The 1994 GATT agreement made clear progress in reducing the potential for use of standards as non-tariff barriers to trade. There remain, however, areas of uncertainty in the agreement that may limit its utility. It is unclear, for example, how capable the new TBT agreement will be in affecting developments in environmental standards. There is great interest at both the national and the international levels in developing new standards and certification systems that link industrial production, trade, and the environment.[32] Whether the WTO and the TBT agreement provide a suitable framework to foster the creation of least trade-distorting environmental standards remains questionable. This is an area that the U.S. government and private sector should certainly monitor carefully.

The TBT agreement also employs vague, nonbinding language committing national governments to harmonization of national standards with international ones, a goal with promise in eliminating barriers to trade in selected product markets.[33] Another area of uncertainty with the new TBT agreement centers on movement toward reciprocity in conformity assessment procedures. Only experience will resolve serious questions as to whether the new agreement can provide a basis to challenge conformity procedures that constitute trade barriers. It is likely, moreover, that problems of interpretation will arise in areas such as laboratory accreditation and quality systems registration in environmental management. It is uncertain, for example, how national governments will interpret Article 6 of the agreement, which requires "whenever possible, that the results of conformity assessment procedures in other Members are accepted, even when those procedures differ from their own, provided they are satisfied that those procedures offer an assurance of conformity."

Efforts by governments to negotiate mutual recognition of conformity assessment procedures provide, therefore, a higher likelihood of reduced barriers than reliance only on the TBT and the new WTO. This raises the importance of quick progress in regional trade forums. Experience gained in the U.S.-EU negotiation on MRAs in conformity assessment, outlined in this chapter, will provide at least some momentum for progress at the multilateral level. Negotiations in a regional framework, such as the early work in APEC, may also prove helpful in building a foundation to mitigate against the development of new conformity assessment barriers in the future.

The next section outlines the institutional processes through which the U.S. government formulates trade policy and sets priorities for removal of barriers to U.S. exports. Unilateral actions authorized under U.S. law designed to retaliate against unreasonable and discriminatory trade practices are discussed. The section concludes with an overview of the history of use of Section 301 by the United States in standards and conformity assessment disputes. This assessment will focus on cases after 1985 and the granting of authority to the USTR to self-initiate 301 investigations.

U.S. TRADE POLICY AND SECTION 301

As noted, the Uruguay Round and the new World Trade Organization represent significant advances in international coverage of issues related to standards and conformity assessment. There remain, however, important work in removing remaining barriers to trade in these areas and a role for unilateral action by the United States. The U.S. trade policy agenda, as it relates to standards and conformity assessment, should seek to aggressively remove barriers and open markets; anticipate future areas where U.S. interests will be affected by emerging standards, testing, and certification systems abroad; and move to take advantage of technical assistance as a tool of export promotion.

This section first provides a brief overview of the U.S. trade policy formation process. In particular, the advisory mechanisms for obtaining private-sector advice on standards-related issues are described. Recent work of the Office of the U.S. Trade Representative in managing trade issues in standards and conformity assessment is outlined. As the primary mechanism through which the United States can address discriminatory foreign trade practices, including removal of non-tariff barriers to U.S. goods embedded in standards or conformity assessment policies, Section 301 of the Trade Act of 1974 is then described.

Overview of U.S. Trade Policy Formation and Implementation

Standards-related trade issues are addressed as part of the broad context in which the U.S. government's international economic policy is formed. To a great extent, this process centers on the duties and mandate of the Office of the United States Trade Representative. In 1934, Congress enacted the Reciprocal Trade

Agreements Act, which delegated to the President authority to negotiate international trade agreements for the reduction of tariffs. The USTR was created by Congress in the Trade Expansion Act of 1962 and given its present name in 1980.[34] In the Trade Act of 1974, presidential negotiating authority was substantially revised and extended, and the USTR was elevated to cabinet-level status.

The USTR has primary responsibility for developing and coordinating the implementation of U.S. international trade policy.[35] The office serves as the principal adviser to the president on trade policy and on the impact of other government policies on international trade. The USTR also has lead responsibility for the conduct of international trade negotiations. This includes negotiations that may relate to standards or conformity assessment systems and international trade. As such, the USTR is the chief representative at multilateral institutions, such as the GATT. USTR also is the lead representative of the U.S. government in dialogue on trade issues at regional forums, such as the APEC forum.[36]

In this role, the USTR conducts bilateral and multilateral negotiations through authority under sections 704 and 734 of the Tariff Act of 1930 and Title I negotiations under the 1974 Trade Act.[37] The office also conducts investigations under Section 301 cases.[38] USTR administers the trade agreements program, including advising on non-tariff barriers, international commodity agreements, and other matters relating to the trade agreements program.

A key function of the USTR is to coordinate trade policy formation with other federal agencies. The USTR is a member of the National Economic Council (NEC) established by President Clinton. The functions of the NEC include coordination of domestic and international economic policy, including foreign trade issues. The NEC has assumed most of the functions and duties of the cabinet-level Trade Policy Committee, the statutory interagency group charged with managing economic policy within the executive branch.

Two subcabinet interagency groups directly support the work of the NEC and USTR on trade policy. It is through these groups that most standards and conformity assessment issues are addressed. The first is the Trade Policy Staff Committee (TPSC). The TPSC is an interagency working group that includes senior-level civil servant representation from U.S. government agencies. The TPSC is supported by more than 60 subcommittees. This includes a Subcommittee on Standards, chaired by USTR, which manages policy coordination and forms U.S. government positions in standards. The second interagency group supporting the USTR is the Trade Policy Review Group (TPRG) at the Deputy USTR or Under Secretary level. When there is a lack of consensus at the TPSC level or in the case of particularly significant policy matters, issues are referred to the TPRG. Matters of especially high priority or controversy at this level are often referred to the Deputies Committee of the NEC for decision and action.[39]

Several agencies, in addition to participating in the work of the NEC and USTR, support standards-related trade policy formation in specific areas of agency mandate or expertise.[40] The Department of Agriculture (USDA) is involved in U.S. trade policy formulation, for example, in areas related to obligations of the United States under the Sanitary and Phytosanitary Standards Code of the GATT. The USDA's Foreign Agricultural Service coordinates work on international issues. The Department of Commerce's (DoC) International Trade Administration (ITA) administers the department's trade responsibilities. This includes representation and assistance on U.S.-EU negotiations on MRAs through the Office of Multilateral Affairs, for example. The U.S. and Foreign Commercial Service (US&FCS) at DoC is also involved in standards-related trade matters through trade promotion functions to assist U.S. exporters, which may include monitoring of foreign standards developments from posts overseas. In addition, the National Trade Data Bank was established and designed to provide a central repository of U.S. government data on international trade and export promotion.

DoC's National Institute of Standards and Technology (NIST) is also involved in assisting the work of the USTR in areas related to international standards. Since NIST's mandate includes primary technical expertise on standards and contact point to the voluntary standards community, staff of the institute are expert on a wide range of matters that impact U.S. commercial relations. NIST represents the Commerce Department, therefore, on the TPSC and serves as the U.S. "inquiry" point for information on proposed regulations that might affect trade for the GATT, as part of U.S. obligations under the GATT Technical Barriers to Trade Code.

The Department of State's Bureau of Economic and Business Affairs has responsibility for formulating and implementing policy regarding foreign economic matters, including standards. The Department of the Treasury's international responsibilities are executed by the Assistant Secretary for International Affairs. The Treasury Department's U.S. Customs Service collects import duties and enforces many laws or regulations relating to international trade. It maintains ties with private business associations, international organizations, and foreign customs services.

Interviews with U.S. government officials and others reveal only limited communication and information transfer between the TPSC subcommittee on standards and primary working groups on domestic standards of the Interagency Committee on Standards Policy chaired by NIST. Aside from overlapping agency representation on these committees, there is no formal link or channel for leveraging the expertise and mandate of both groups. There is clearly the need for enhanced communication and coordination between interagency committees with trade and domestic standards policy functions, as well as communication between these groups and the private sector. In particular, the work of the USTR in standards and conformity assessment requires the full and active cooperation of other executive branch agencies to support U.S. trade objectives. This is especially true as the USTR leads government efforts in new areas of policy formation and negotiation, such as emerging environmental management systems standards. This applies to the USTR's work in fulfilling not only multilateral objectives, but also standards issues addressed in regional and bilateral negotiations outlined in this report.

Private-Sector Advisory Mechanisms

In 1974, Congress established a private-sector advisory committee system to ensure that U.S. trade policy and trade negotiation objectives adequately reflect U.S. commercial and economic interests.[41] The USTR manages the advisory committees in cooperation with executive branch departments.[42] There are 38 advisory committees in the system. The Advisory Committee for Trade Policy and Negotiations (ACTPN) is the senior level of the three-tiered advisory system. It is a 45-member presidentially appointed body (two-year terms) that provides guidance on trade policy matters, including trade agreements and negotiations.

The seven policy advisory committees make up the second tier. Members are appointed by the USTR in conjunction with one or several secretaries.[43] They are the Services, Investment, Intergovernmental, Industry, Agriculture, Labor, and Defense Policy Advisory Committees. Their role it is to advise the government of the impact of various trade measures on specific sectors (e.g., industry and agriculture) of the economy.

Most standards-related trade issues are informed through a third tier in the private advisory system, composed of 30 sector, functional, and technical advisory committees. Members of these groups are appointed by the USTR and the secretaries of Commerce and Agriculture. Sectoral or technical committees provide technical information and advice on trade issues affecting specific sectors. The functional advisory committees provide advice on customs, standards, and intellectual property issues.[44] The Industry Functional Advisory Committee (IFAC) on Standards for Trade Policy Matters serves as the formal mechanism for industry advice on standards and trade. It meets only at irregular intervals or at the call of USTR or the Secretary of Commerce.

Removing Standards-Related Trade Barriers: Section 301

Future U.S. trade interests, as they relate to developments in international standards, will be served not only through multilateral institutions and agreement, but also through unilateral action by our government. Section 301 of the Trade Act of 1974, as amended, provides a useful tool to address barriers to trade in standards and conformity assessment systems.[45]

Section 301 supports petitions by individuals or commercial interests to enforce U.S. rights under international trade agreements, as well. The USTR may also self-initiate a 301 action under its own authority. Under the law, USTR must determine within 45 days whether or not a petition for action under Section 301 is justifiable and should be pursued.[46] The USTR contacts foreign trade partners and requests discussions for dispute settlement. If trading partners refuse to participate in negotiations, the USTR may make recommendations to the president to proceed with unilateral retaliatory measures.[47]

In addition to enforcing U.S. rights in international law, Section 301 is targeted at responding to unreasonable, unjustifiable, or discriminatory trade practices that serve to restrict U.S. trade.[48] It is the primary means of unilateral action to open overseas markets for U.S. goods through the removal of unfair foreign trade practices.[49] This includes barriers to trade reflected in standards or conformity assessment systems, although it has not been employed in any systematic manner to address these disputes.

The use of trade policy tools, such as Section 301, has clearly not always been uniformly effective in meeting U.S. objectives. Recent analyses of the use of Section 301 provide evidence, however, that it can in many cases serve to support trade liberalization and reduce standards-related barriers to U.S. exports.[50] Overall, the law is a relatively forceful tool of unilateral action. In one study of 72 cases initiated under Section 301 from January 1975 to June 1994, U.S. negotiating objectives were met in 35 of 72 cases. This contrasts favorably with a more limited, 33 percent success rate found in an analysis of the outcomes of 120 foreign policy sanction cases involving the United States.[51] There are a number of general characteristics of standards and conformity assessment cases that are consistent with the broad outlines of successful action under Section 301. For example, there appears to be a higher rate of success in 301 actions involving relatively more transparent border measures against imports. This might reasonably extend to at least some of the more visible actions by foreign governments to (1) forbid acceptance of certifications of laboratory accreditations, (2) deny documented test results transmitted at customs, and (3) raise other serious barriers in conformity assessment regulations.

What are other characteristics of the successful use of 301 that should be considered in addressing future barriers to trade in foreign standards and conformity assessment policies? In general, the United States has been successful when the target country is more highly dependent on the U.S. market for exports (77 percent of the total cases). The use of Section 301 has also been relatively more effective when a case is *self-initiated by USTR.* This has been particularly true in negotiations with the European Union. The success rate for cases launched against Europe reached 75 percent, when initiated by the USTR.

The case of wood products imports to Japan serves as a concrete example of the utility of Section 301 in standards-related cases. USTR began Super 301 investigation in 1989, in part at the urging of the National Forest Products Association.[52] Negotiations led to several changes in Japanese policy including the acceptance of foreign testing and certification and the use of less restrictive and more performance-oriented standards. These processes are overseen by standing U.S.-Japanese technical committees. The standards-related part of the complaint has been addressed and resolved without trade retaliation.

MUTUAL RECOGNITION AGREEMENTS

The preceding sections of this chapter discussed mechanisms for removing trade barriers through multilateral action in the GATT's World Trade Organization and through unilateral retaliation against unfair foreign trade practices, as authorized by U.S. trade law. As awareness spreads among U.S. industry of these mechanisms, particularly the new dispute resolution processes in the GATT, they have the potential to contribute significantly to the reduction of standards and conformity assessment systems that discriminate against U.S. exports.

These approaches, however, involve costs and can be highly confrontational. The threat of U.S. trade retaliation, whether unilateral or sanctioned by the GATT, should not be the only means of persuasion available to open foreign markets to international goods. Cooperative negotiations with U.S. trading partners aimed at gaining mutual benefit also hold great potential for facilitating trade and leveraging economic benefits. In particular, negotiations to achieve mutual acceptance of conformity assessment procedures—testing, certification, accreditation, and quality system registration—can reduce existing barriers to trade. There must, obviously be clear oversight and assurance of competence in any agreement on mutual recognition. Under these conditions, however, an MRA can preempt the development of new barriers that may arise as nations develop increasingly complex infrastructures for testing and certifying tradable goods and services.

There are other indications that MRAs and progress toward cooperation across national boundaries can be beneficial to U.S. and international interests. A detailed analysis of the progress made through the North American Free Trade Agreement (NAFTA) among the United States, Canada, and Mexico in standards harmonization and conformity assessment is beyond the scope of this report. Continued research and analysis, however, of standards issues related to the NAFTA are important. In general, the NAFTA provisions on standards, testing, and certification indicate that regional trade agreements can be reached that both ensure continued high levels of health, safety, and environmental standards and move toward harmonization and mutual recognition between nations with differing levels of economic development.[53]

Background: Product Approval in the European Union

In its request for this study, Congress asked for an appraisal of changes in the standards and conformity assessment systems of the European Union and ways for U.S. exporters to respond to the challenges they present. The European Union is one of the largest export markets for U.S. firms, accounting for 20.9 percent of U.S. exports in 1993.[54] Many changes have occurred in European trade policies and systems as part of the drive to establish a common internal economic market. To the extent that these changes succeed in removing trade barriers among European nations, they have the potential also to facilitate U.S. exports to Europe as a whole. Consolidation of multiple European markets with formerly separate regulations and product certification requirements reduces the burden on U.S. exporters of gaining information about, and meeting, multiple sets of criteria.

New European product approval regulations, however, have erected a potentially significant trade barrier around the European market as a whole.[55] There are two components to these new regulations: product standards and systems for assessing conformity to those standards. In 1985, the European Commission addressed the former issue with its "New Approach to Technical Harmonization." The New Approach abandoned the pursuit of lengthy, detailed negotiations to harmonize (rewrite as equivalent) the national technical regulations of different member states—a process that had proved extremely slow and cumbersome. Instead, the Commission would issue directives listing less detailed, "essential requirements" for safety that regulated products would have to meet.[56] These directives would set a required level of safety without dictating means for achieving it. Detailed standards for meeting the essential requirements would be allowed to vary among European states until new, pan-European standards could be written. A product meeting one country's standards, however, could not be denied access to any other European market, even if it did not meet the detailed standards of the destination market's national regulations.

The Commission delegated to the private sector the writing of new technical standards linked to EU-wide essential product requirements. This system in which the private sector leads in standards development now mirrors, to a great extent, the U.S. model of standards development. Pan-European technical standards are now being developed, under contract with the Commission, by three private standards-developing organizations (see Box 4-1). These are the European Commission for Standardization (CEN), the European Commission for Electrotechnical Standardization (CENELEC), and the European Telecommunications Standards Institute (ETSI). The members of CEN and CENELEC are the national standards bodies of Europe, while ETSI includes national telecommunication agencies, manufacturers, and industry associations.

Standards developed by these organizations play a central role in determining what products may be marketed in Europe. CEN, CENELEC, and ETSI standards are not the only standards the EU will accept as meeting the essential product directives. Products complying with other standards are acceptable, as long as the alternative standards also meet the EU essential requirements. The burden of proof in such cases, however, is on manufacturers. For this reason, product approval is easier to obtain through compliance with the CEN/CENELEC/ETSI standards. Participation in their standards-writing work is therefore of clear benefit to firms that market regulated products in Europe. Unlike most U.S. standards-developing organizations, CEN, CENELEC, and ETSI are not open to foreign participants. As outlined in Box 4-1, however, there are several avenues for U.S. firms to influence these organizations' standards work.

BOX 4-1
EUROPEAN REGIONAL STANDARDS-DEVELOPING ORGANIZATIONS

CEN: Comite Europeen de Normalisation (European Committee for Standardization). Based in Brussels, CEN has a membership consisting of the national standards-writing organizations of 18 European countries—the members of the European Union and European Free Trade Area (EFTA). CEN develops voluntary European Standards in all product sectors excluding electrical standards covered by CENELEC. With funding from the European Commission, CEN also writes standards to meet the "essential requirements" for product safety mandated in EU product directives. The standards work program is directed by seven technical sector boards covering building and civil engineering; mechanical engineering; health care; workplace safety; heating and cooling; transport and packaging; and information technology. CEN maintained 1,134 active standards and 265 technical committees as of October 1993.

CENELEC: Comite Europeen de Normalisation Electrotechnique (European Committee for Electrotechnical Standardization). Like CEN, CENELEC is based in Brussels and has 18 European standards bodies (national electrotechnical committees) as members. CENELEC develops European Standards for electrotechnology, which includes areas such as consumer electronics, electric power generation, electromagnetic compatibility, and information technology. International standards developed by the International Electrotechnical Commission are the basis for 89 percent of CENELEC standards. CENELEC also develops standards meeting EU product directives, with funding from the European Commission. Approximately 35,000 technical experts participate in CENELEC standards-writing committees.

ETSI: European Telecommunications Standards Institute. Based in Sophia Antipolis, France, ETSI is not part of the CEN/CENELEC structure, but has a cooperation agreement with those organizations. Membership is composed of the public telecommunications administrations of EU and EFTA nations, as well as manufacturers and trade associations. ETSI develops European Telecommunications Standards, which may be adopted as mandatory by European national telecommunications systems. ETSI has adopted due process procedures that require less consensus than CEN and CENELEC, in an effort to hasten the standards development process. In 1992, ETSI published 184 standards and technical reports.

As of April 1992, European standards organizations were at work on approximately 2,000 standards. This work was concentrated in the construction, aerospace, information technology, and machinery sectors, as well as public procurement. To the extent that European standards vary without justification from international standards in equivalent sectors, they represent barriers to imports from outside Europe. This danger is reduced, however, by CEN's and CENELEC's pledge to defer writing standards when International Organization for Standardization (ISO) and IEC standards exist or are under development in the same product sectors. This pledge underscores the importance of U.S. industry's participation in ISO/IEC technical committee work as a way to influence standards-setting in the European market. This strategy has been pursued successfully, for example, by the U.S. medical devices and construction equipment industries.

In addition to influencing European standards through international (ISO/IEC) technical work, U.S. manufacturers have several other routes of access to CEN, CENELEC, and ETSI. These include ongoing consultations and information exchange between ANSI and European standards organizations, which include ANSI-coordinated bilateral discussions held in Europe each year between industry and government officials; comments on official notices of new European standards work programs; and direct participation in European standards work by European subsidiaries of U.S. corporations. Direct participation in CEN, CENELEC, and ETSI standards development is prohibited, however, for U.S. firms without a substantial European presence. The converse is not the case; foreign firms have open access to participate in the U.S. voluntary consensus standards system.

SOURCES:

U.S. International Trade Commission (USITC), *The Effects of Greater Economic Integration Within the European Community on the United States: Fifth Followup Report*. Washington, D.C.: USITC, April 1993.

CEN, "What is CEN?", brochure, 1993.

CENELEC, "CENELEC: Electrotechnical Standards for Europe", brochure, n.d.

ETSI, *Annual Report 1992*. Sophia Antipolis: ETSI, 1992.

Michael Miller, President, Association for the Advancement of Medical Instrumentation, presentation to the Conference on New Developments in International Standards and Global Trade, Washington, D.C., March 30, 1994.

American National Standards Institute (ANSI), *The U.S. Voluntary Standardization System: Meeting the Global Challenge*, 2nd ed. New York: ANSI, 1993.

Maureen Breitenberg, ed., *Directory of European Regional Standards-Related Organizations*, NIST Special Publication 795. Gaithersburg, Md.: National Institute of Standards and Technology, 1990.

On the whole, limits on U.S. input to European standards organizations do not appear to be a serious restriction to trade.[57] Continued monitoring, however, is necessary to forestall problems in this area.

To a greater extent than standards development, the most serious potential barrier to EU imports from the United States and other non-European countries is in the area of conformity assessment. In 1990, the EU's New Approach to harmonization of standards was followed by a set of changes in mechanisms for proving product conformity to standards. The European Commission instituted these changes as the "Global Approach to Testing and Certification." This is a framework for showing compliance to the EU New Approach product directives that relies on mutual recognition of conformity assessment systems within Europe. The framework consists of technical rules for conformity assessment procedures outlined by the European Commission. To prove compliance with product directives, manufacturers must use one or more specified conformity assessment procedures, including self-certification (manufacturer's declaration of conformity), third party testing and certification, and quality system registration.[58]

In the product sectors in which third-party product testing, certification, or quality system registration is required by law, this approval may be granted only by organizations designated, or "notified," to the Commission by the member states as technically competent. These "notified bodies"—and no others—may grant final approval of products for the European market. Certified products are identified with the newly introduced, European "CE Mark." Products bearing this mark are certified as meeting the EU's essential product directives and may circulate freely throughout Europe. Products without the CE Mark may not be marketed in Europe at all.

Although the new European testing and certification system will facilitate the flow of trade within Europe, it has raised potentially significant barriers to U.S. products at the conformity assessment level. The requirement that final assessments be performed by European "notified bodies" raises the costs of testing and certification to U.S. manufacturers in many sectors. As outlined in the next section, this issue is now the subject of ongoing talks between the United States and the EU.

U.S.-European Union MRA Negotiations

In early 1994, the United States entered into negotiations with the EU on mutual recognition of conformity assessment systems. The goal of the negotiations is a framework under which European and U.S. exports are facilitated through mutual recognition by both the U.S. and Europe of third-party product test results, inspections, and certifications. The U.S. goal is to increase U.S. exporters' access to European markets and overcome barriers imposed by the new EU product conformity requirements. EU negotiators seek improved access for European products to the U.S. market. The EU expectations in the negotiations are that through an MRA with the United States, it will gain formal U.S. government assurance that U.S. entities within an MRA are competent to perform "essential" services in inspection and certification, as required by EU mandates. The EU, therefore, through an MRA with the United States could receive formal assurance that is lacking under current conditions. The EU also expects that in an MRA, the United States will accept European entities as competent to test to U.S. government requirements. They will gain this without the need for multiple accreditation by the U.S. government when already accredited by the EU.

The negotiations are intended to produce a series of bilateral, U.S.-EU mutual recognition agreements, each covering a specific product sector regulated by the EU. Mutual recognition by national governments of testing data, laboratory accreditation, and product certifications against specific standards represents significant potential for increased trade. The talks illustrate the potential economic benefit of MRAs and offer insight into the possibilities for agreements with other nations and regional associations.

U.S. negotiators—led by the Office of the USTR and the Department of Commerce, with participation and advice from private industry—are seeking to outline rules under which U.S. manufacturers could obtain certification to EU requirements through U.S. owned laboratories and certifiers. Under mutual recognition, European manufacturers would also gain greater access to U.S. certification, including increased opportunities for conducting laboratory tests within Europe for products regulated in the United States.[59] This would obviously involve acceptance by the United States of work performed in Europe by European entities. MRAs are under discussion in 11 sectors: information technology; telecommunications products attached to public networks; medical devices; electrical safety; electromagnetic interference; pharmaceuticals; pressure equipment; road safety equipment; lawn mowers; recreational boats; and personal protective equipment such as helmets.[60]

There are a number of benefits for U.S. producers associated with MRAs in these sectors. At present, U.S. firms have three avenues for obtaining required third-party certifications for the EU market. They can ship product samples to Europe for testing and certification by a European notified body and pay expenses for European inspectors to inspect their plants in the United States. They can have testing and certification performed by one of a growing number of U.S. subsidiaries of European laboratories. Or in some product sectors, they can have testing performed by a U.S. laboratory that subcontracts to a European certifier. In the latter case, the U.S. laboratory performs required tests, then forwards the test data to a European laboratory for evaluation and final approval for the CE Mark.

All three of these avenues exclude U.S. testing laboratories from the final stage of product certification—the judgment of test results and approval of the product. The seriousness of this burden varies for different product sectors. In some sectors, such as the gas appliances industry, subcontracting to European laboratories has proved sufficient to meet U.S. firms' market access needs.[61] In others, subcontracting is less satisfactory. A key example is testing of electronic products for compliance with EU directives on electromagnetic interference. Under subcontracting arrangements, U.S. laboratories are not allowed by EU regulators to exercise "engineering judgment" and must therefore perform redundant, additional tests that European laboratories are not required to perform. As a result, U.S. laboratories must charge higher testing fees to U.S. manufacturers.[62]

In summary, the present avenues for U.S. firms to obtain EU product approval represent a barrier to U.S. exports. They frequently entail such unnecessary costs as redundant testing of products already tested against U.S. standards; shipping product designs and prototypes overseas for testing; and transportation of foreign engineers to the United States for factory inspections. Mutual recognition of conformity assessment between the United States and the European Union would permit U.S. manufacturers to obtain required European certification at the U.S. location of production. It would open the market for third-party certification services to U.S. laboratories, resulting in greater competition and lower costs to U.S. manufacturers. It would also directly benefit the U.S. independent testing industry. As discussed in Chapter 3, this is a rapidly growing, $10.5 billion industry.

There are several obstacles to successful completion of the U.S.-EU talks.[63] As a result of the significant differences in the two sides' conformity assessment systems, negotiations have involved time-intensive exchange of background information and sector-specific data. Differences in the structure and operation of the systems create difficulty in developing means for full, reciprocal access to conformity assessment procedures. Apart from a few exceptions, such as drug certification and automobile emissions testing, the United States relies much more than the EU on manufacturer's declaration for regulatory approval.[64] The EU system centers primarily on third-party conformity assessment in regulated sectors. These third-party procedures include testing and certification by independent laboratories, as well as quality system audit and approval by third-party registrars. As noted previously, only "notified bodies"—those designated as technically competent by their national government—may perform these third-party services.

At present, there is no mechanism for any non-European organization to become accepted by the EU as a notified body. Under an MRA, U.S. organizations could become notified bodies and perform testing and certification of exports to the EU. The EU has indicated that any mutual recognition will require some form of U.S. government involvement in guaranteeing the competence of private U.S. conformity assessment organizations before they will be accepted by EU regulatory authorities. As a first step toward creating such a mechanism, in 1994, the National Institute of Standards and Technology created the National Voluntary Conformity Assessment Systems Evaluation (NVCASE) program. Under NVCASE, NIST will recognize U.S. accreditation programs in the United States that are competent to evaluate U.S. conformity assessment organizations.

To minimize government involvement in the private U.S. conformity assessment system, NIST intends to restrict NVCASE primarily to the *recognition* level of activity.[65] (See Figure 3-1 for the levels of conformity assessment in the U.S. system.) NVCASE is designed to evaluate and officially recognize the competence of accreditation bodies (accreditors) who, in turn, accredit testing laboratories, certifiers, and quality system registrars. NVCASE will not directly *accredit* testing laboratories or certifiers, unless there exists no private-sector accreditation program in a particular industry sector and firms in that sector ask NIST to set up a program. Recognition by NVCASE confers U.S. government approval on all testing, certification, and quality system registration procedures performed by parties who are, in turn, accredited by NVCASE-recognized accreditors. The

NVCASE evaluation criteria are to be designed on a case-by-case basis, but NIST has confirmed that accreditors will be required to follow internationally recognized procedures, such as those developed by the International Organization for Standardization (ISO).

In meetings in Washington, D.C., in June 1994, EU negotiators gave preliminary indication that the NVCASE program would prove sufficient, under a broader MRA, for the EU to accept competence of U.S. organizations as notified bodies.[66] This government contact point, at the recognition level, is important as a way to overcome the many differences between Europe and the United States at the accreditation and assessment levels. With government oversight and assurance at the top, private entities will perform the specific tests necessary for product approval. In fact, some European states—notably, the Netherlands—already rely on a similar system for designating notified bodies. The Dutch government recognizes a private organization, the Raad vor Certificaat (Certification Council), to accredit laboratories, certifiers, and quality system registrars for regulatory conformity assessment activities.[67]

The European goal for mutual recognition is to gain greater access to the U.S. market for European exports. European negotiators have expressed concern about the complexity of the U.S. conformity assessment system, with public components at the national, state, and local levels and a variety of private certification systems. They note, for example, the lack of a U.S. national or North American mark for entry into the United States, Canada, and Mexico analogous to the European CE Mark.[68] A U.S. national mark would, theoretically, guarantee free movement of products throughout 50 states and hundreds of localities, many with unique regulations and certification schemes. U.S. product liability law and its consequences for laboratories accustomed to operating in other legal environments represent another concern cited by European negotiators.[69]

As discussed in detail in Chapter 3, the U.S. conformity assessment system is complex. Its complexity imposes many obstacles to commerce within the United States, and this is true for both domestic U.S. firms and those based in Europe and Asia wishing to serve the U.S. market. It does not, however, in a legal, government-mandated manner, pose a discriminatory international trade barrier, as noted in several recent reports of the EU on trade and investment with the United States.[70] The U.S. conformity assessment system, as well as associated product liability practices, is as complex for U.S. firms to grapple with as it is for those in Europe. In addition, unlike the European system, there is no U.S. policy excluding foreign laboratories from participation. Foreign laboratories can and do obtain accreditation within the U.S. system, through both public accreditors (such as NIST's National Voluntary Laboratory Accreditation Program) and private ones (such as American Association for Laboratory Accreditation).[71] Necessary measures to improve the U.S. system, which are proposed to help streamline and promote U.S. economic efficiency in the national interest, such as those discussed in Chapter 3 and recommended in Chapter 5, will also serve to mitigate against claims by European and other foreign manufacturers that they confront unnecessary burdens in the U.S. market.

Although there is clearly duplication among state, local, and federal government rules and regulations regarding conformity assessment procedures in the U.S. system, the European Union's approach to product approval is a discriminatory trade barrier. EU policies *mandate* product certification by European notified bodies—selected testing laboratories, certifiers, and public agencies that can only be designated by national governments in the European Union. The system imposes duplicate testing costs on U.S. manufacturers as they seek to compete in the European market. It also prevents U.S. testing laboratories, except those acting as subcontractors to European laboratories, from carrying out all of the functions needed to grant use of the CE Mark on regulated products sold in Europe.

To overcome this barrier to U.S. products and promote mutual expansion of international trade, continued negotiations by the USTR and the Department of Commerce on MRAs with Europe are highly desirable. U.S.-EU mutual recognition has the potential to create strong expansion in international trade opportunities for U.S. and European firms alike. It will also set a strong, favorable precedent for MRAs with other U.S. trading partners. If European negotiators, however, refuse to accept the competence of the U.S. conformity assessment system and provide fair access to European markets by signing MRAs within a reasonable period of time, it will be necessary for the USTR to proceed to the next step in removing European barriers to U.S. exports—initiating trade reprisals under the authority of Section 301.

Mutual Recognition Agreements: APEC

The MRA talks between Europe and the United States offer one model of how to structure dialogue on conformity assessment. Success in opening market access through these talks will set a valuable precedent for similar agreements with other U.S. trading partners. As the text of the Uruguay Round Technical Barriers to Trade Agreement acknowledges, mutual recognition of conformity assessment is important to future global trade expansion. MRAs are an important avenue, not only for overcoming existing barriers to trade, but also for averting the emergence of new barriers as nations develop increasingly complex conformity assessment systems.

Beyond the European Union, there are other priority regions where streamlined conformity assessment through MRAs would benefit U.S. trade and global export expansion. This is particularly true in Asia and Latin America. In 1994, President Clinton identified expanding U.S. exports in emerging Asian and Latin American markets as a top economic policy priority.[72] As Table 4-4 illustrates, the most rapid growth of regional markets for U.S. exports over the next 10 years will be in Asia and Latin America. The president's 1994 report to the Congress

TABLE 4-4 — Projected Growth of Regional Trading Partners

REGIONAL MARKET	GDP,[a] 1993 (billions of U.S. dollars)	Share of World Total, 1993 (%)	Estimated Growth Rate,[b] 1993-2000 (%)	Projected GDP,[a] 2003 (billions of U.S. dollars)	Share of World Total, 2003 (%)
Latin America, i.e., Mexico	1,757	7.7	5.2	2,917	8.9
Mexico	661	2.9	4.7	1,046	3.2
East Asia, i.e., Japan	5,027	22.2	6.2	9,174	27.9
Japan	2,548	11.2	2.4	3,230	9.8
Pacific	370	1.6	2.4	469	1.4
Canada	617	2.7	3.1	837	2.5
Western Europe	6,538	28.8	2.5	8,369	25.4
Middle East	1,174	5.2	3.9	1,721	5.2
Other Asia	1,598	7.0	2.5	2,046	6.2
Africa	654	2.9	2.5	837	2.5
Eastern Europe	1,751	7.7	2.5	2,241	6.8
Total	22,695			32,888	

[a] Purchasing Power Parity estimates
[b] Data Resources, Inc.

SOURCE: Clinton, William J. *Report to the Congress on Recommendations on Future Free Trade Area Negotiations, July 1, 1994.* Adapted from Table #2.

on future free trade area negotiations projected East Asian markets, excluding Japan, to grow at an annual rate of 6.2 percent through 2003. Latin American markets, apart from Mexico, will grow at a 5.2 percent annual rate.[73] With Japan and Mexico included, East Asia and Latin America will represent almost 50 percent of the world's combined GDP by 2003.

Table 4-5 further underscores the importance of Asian and Latin American markets to U.S. export success. Since 1985, the rate of growth in U.S. exports to Asia, excluding Japan, has been almost double that of exports to Europe or Canada. Growth of U.S. merchandise exports to selected Asian and Latin American countries is presented in Tables 4-6 and 4-7. China, Taiwan, Singapore, Hong Kong, South Korea, and Mexico have all grown rapidly as substantial markets for U.S. goods. Indonesia, Argentina, Chile, and the Dominican Republic, meanwhile, are fast emerging as significant future U.S. trading partners. The emerging markets of Asia and Latin America clearly represent a vast potential for future U.S. export expansion. To the extent that standards and conformity assess-

TABLE 4-5 — U.S. Exports to Regional Partners—Changes
since 1985 (billions of U.S. dollars)

REGIONAL PARTNER	1985	1993	CHANGE (%)
Latin America	31.0	72.9	135.2
East Asia-Pacific, i.e., Japan	31.4	87.5	178.7
Japan	22.6	47.9	111.9
Canada	53.3	100.2	88.0
European Union	49.1	97.1	97.8
Rest of the world	31.4	58.9	87.6
Total	218.8	464.5	112.3

SOURCE: Clinton, William J. *Report to the Congress on Recommendations on Future Free Trade Area Negotiations, July 1, 1994.* Adapted from Table #1.

TABLE 4-6 — Growth of U.S. Merchandise Exports to Selected APEC Members
(millions of U.S. dollars)

COUNTRY	1986	1993	CHANGE (%)
Australia	5,551	8,272	49.0
China	3,106	8,767	182.3
Hong Kong	3,030	9,873	225.8
Indonesia	946	2,770	192.8
Japan	26,882	47,949	78.4
Malaysia	1,730	6,064	250.5
Singapore	3,380	11,676	245.4
South Korea	6,355	14,776	132.5
Taiwan	5,524	16,250	194.2
Total	56,504	114,721	103.0

SOURCE: Clinton, William J. *Report to the Congress on Recommendations on Future Free Trade Area Negotiations, July 1, 1994.* Adapted from Table #4.

TABLE 4-7 — Growth of U.S. Merchandise Exports to Selected Latin American and Caribbean Countries (millions of U.S. dollars)

COUNTRY	1986	1993	CHANGE (%)
Argentina	944	3,771	299.5
Brazil	3,885	6,045	55.6
Chile	823	2,605	216.5
Colombia	1,319	3,229	144.8
Costa Rica	483	1,547	220.3
Dominican Republic	921	2,350	155.2
Guatemala	400	1,310	227.5
Mexico	12,392	41,635	236.0
Paraguay	171	521	204.7
Venezuela	3,141	4,599	46.4
Total	24,479	67,612	176.2

SOURCE: Clinton, William J. *Report to the Congress on Recommendations on Future Free Trade Area Negotiations, July 1, 1994.* Adapted from Table #3.

ment agreements can be tailored to promote trade expansion in this area, significant U.S. economic benefit will result.

Progress toward this goal has already begun in the Asian region. In 1994, preliminary discussions on mutual recognition of conformity assessment began within the APEC forum.[74] APEC consists of most of the principal economies bordering the Pacific Ocean: the United States, Canada, Japan, South Korea, Taiwan, Hong Kong, Singapore, China, Australia, New Zealand, and the members of the Association of Southeast Asian Nations (Malaysia, Indonesia, the Philippines, Thailand, Brunei, and Myanmar). Chile joined APEC in 1994. Together, these countries account for approximately 40 percent of global economic activity. That share is increasing rapidly.

U.S. trade relations within the Asia Pacific region, as in Europe, are influenced to a significant degree by standards and conformity assessment concerns. As noted earlier in this chapter, the USTR has identified significant technical trade barriers in Asian markets. U.S. trade retaliation has achieved some success in removing these barriers. Mutual recognition of conformity assessment mechanisms, however, has the potential to further the goals of free and open trade in a constructive and less confrontational manner in regional discussions, such as those supported through APEC.

Early preparatory work for possible mutual recognition discussions began in an informal APEC working group on standards and conformity assessment. This process was further advanced through the endorsement of the principle of an APEC MRA by the Eminent Persons Group created to advise the APEC member nations and leaders on achieving more open trade in APEC.[75] The APEC Leaders Meeting in Jakarta, Indonesia in November 1994 endorsed the broad concept of MRAs among APEC members, as part of its declaration to begin the process of region-wide open trade by the year 2020.[76]

The declaration on an APEC standards and conformity assessment framework agreed to in Jakarta outlines several key principles for such mutual recognition. The objectives of the framework include both reducing barriers to trade within the region and promoting "the further development of open regionalism and market-driven economic interdependence in the Asia Pacific region."[77] Noting the benefits to be gained from reducing "unnecessary costs and time-consuming delays associated with duplicate testing of products," the draft states that, "the development of broader networks of mutual recognition arrangements through the region will be a key objective of APEC's work program."

The key elements of an MRA, identified in the draft submission to the APEC Leaders Meeting in Jakarta are presented in Box 4-2. These include the need for a clear definition of the scope of testing and certification procedures mutually accepted by the parties to the MRA; criteria for identifying competent, acceptable laboratories and certifiers in each country; provisions for information exchange, joint monitoring, and dispute resolution; and a commitment by government authorities in each country to oversee the performance of conformity assessment organizations and, if necessary, terminate their accreditation if they fail to maintain technical competence.

Achieving MRAs in APEC will involve a detailed and complex set of policy planning exercises over the next several years. Further analysis will be necessary to support these negotiations. There are significant differences among APEC nations by level of industrialization, nature of government oversight of standards and certification systems, and state of technical infrastructure capacities. As the experience of U.S.-EU negotiations demonstrates, differences in conformity assessment systems complicate the prospects of reaching full mutual recognition, at least in the short term. Within APEC, a more modest approach is likely needed than in the U.S.-EU dialogues. This approach should start with the development of specific proposals and a framework for information exchanges to build confidence in testing data throughout the region, for example, as a step toward acceptance of product certifications.

In addition, technical assistance from the industrialized nations of APEC will be necessary to enable the developing member nations to create transparent, competent systems for establishing product regulations and assessing conformity

BOX 4-2
ELEMENTS OF A MODEL MUTUAL
RECOGNITION AGREEMENT

It is suggested that the following are likely to be desirable key elements in a government-to-government mutual recognition agreement, whether it be a broad overarching agreement or an agreement that is sector specific:

• An undertaking on the behalf of each party to grant mutual acceptance of reports and/or certificates and, where applicable, marks of conformity issued by approved bodies of the other; these reports and certificates would have the same validity as the equivalent documents issued by bodies in the recognizing country.
• A clear definition of the coverage of the agreement, whether it be in terms of regulatory regime or types of products and/or services, the relevant mandatory third-party assessment requirements and procedures need to be listed.
• An agreed and clearly specified process or criteria for determining what would be regarded as acceptable conformity assessment bodies. Those criteria would desirably include a requirement that testing and inspection bodies meet the rules in relevant ISO/IEC guides such as 25 (currently under revision), 39 (a revision of which is expected shortly), and 43. It is also desirable that the criteria include a requirement that to be eligible, a conformity assessment body must be accredited by an accreditation body that is regarded as acceptable by both parties and must be able to demonstrate to the accreditation body that it can assess conformity to the requirements of the other party. The agreement, or an annex thereto, will need to list the bodies that are jointly agreed as being competent to undertake conformity assessment.
• An undertaking by each party that the government authorities who have the responsibility for determining and listing the agreed conformity assessment bodies will have the power and competence to address any problems that arise and to remove, if necessary, unacceptable bodies from the list.
• Established mechanisms for ensuring that the conformity assessment bodies listed by each party meet, and continue to meet, the criteria required by the other party; these would desirably include some form of intercomparison procedures and regular audits.
• An undertaking by each party to permit, at the request of the other party, checks to be carried out to ensure that the listed conformity assessment bodies comply with the criteria that have been set.
• An undertaking to supply all relevant information to the other party including notice of any changes to legislative, regulatory, and administrative provisions.
• Provisions for joint monitoring and administration of the agreement and for dispute resolution.
• It may also be appropriate to include provisions for subsequent broadening of the scope of the agreement with the consent of both parties.

SOURCE: "Development of a Model Mutual Recognition Agreement on Conformity Assessment," Document III.6.3.3.1, submitted by Australia to the Committee on Trade and Investment, Asia-Pacific Economic Cooperation Sixth Ministerial Meeting, Indonesia, 1994.

to those regulations. As discussed in the following section of this chapter, technical assistance related to standards and to conformity assessment benefits both recipient countries, by modernizing their economic infrastructure, and donor countries, by expanding their potential export markets.

U.S. EXPORT PROMOTION

The following material moves beyond current trade developments to identify links between standards and opportunities for U.S. export expansion. The greatest potential for trade expansion in future years is represented by the dynamic economic growth in key emerging markets. As Tables 4-4 through 4-7 illustrate, emerging Asian and Latin American markets represent especially promising opportunities for U.S. firms in new global markets.

U.S. Export Promotion Policy

Export promotion activities are, along with trade barrier reduction, a central component of U.S. trade policy. As discussed earlier in this chapter, expansion of U.S. export opportunities in emerging markets contributes in critically important ways to the long-term potential for U.S. economic advance. A variety of U.S. agencies conduct programs to promote exports. These programs incorporate mechanisms such as the following: (1) direct funding to U.S. firms and nongovernment organizations for technical assistance projects within developing countries; (2) low-interest loans to developing countries to support importation of goods and services, particularly from the United States; (3) information dissemination to U.S. firms about export opportunities, through a variety of publications and the computerized National Trade Data Bank; and (4) trade fairs promoting U.S. products in key export markets.

Eight U.S. government agencies and quasi-governmental organizations are principally engaged in export promotion activities.[78] These are the Department of Agriculture; the Department of Commerce, including the U.S. and Foreign Commercial Service and the International Trade Administration; the Department of Energy; the Agency for International Development; Eximbank; the Overseas Private Investment Corporation; the Small Business Administration; and the Trade and Development Agency, which provides grants for project feasibility studies. These and other executive branch agencies coordinate U.S. export promotion policies through the interagency Trade Promotion Coordinating Committee, founded in May 1990.

Although U.S. export promotion accounted for about $2.7 billion in spending as recently as fiscal year (FY) 1991, these activities have not been linked effectively to the nation's comparative advantages in advanced-technology exports. For example, agriculture export promotion programs accounted for $2.0 billion of the $2.7 billion in spending, although agriculture accounts for only about 10 percent of total U.S. exports.[79] Enhancement of long-term opportunities for U.S. exports calls for innovative, new programs aimed at strengthening the technological infrastructure of developing countries that represent emerging export markets. These programs will create increased demand for goods and services, which U.S. exporters are well prepared to meet.

Technical assistance related to standards and to the technical infrastructure necessary to test and certify products represents a vital opportunity for promotion of U.S. exports. Product standards, systems for developing them, and national frameworks for assessing conformity are all being put rapidly into place in newly emerging market economies worldwide. This is particularly the case in Asia. Technical standards assistance and growing export opportunities are linked in two key ways. First, active promotion of U.S. standards and U.S.-developed international standards has the potential to facilitate U.S. industry's access to key markets, to the extent that these standards are adopted as requirements for product acceptance. In addition to promoting specific U.S. industrial products or technologies, technical assistance should foster the use of international standards. International standards provide a developing country with the greatest possible range of choice of goods and services, and they create a level playing field for exports from the United States, Europe, Japan, and other industrialized regions.

Second, as discussed previously in this chapter, there is a strong link between continued liberalization in global trade and rising wealth and standards of living in all nations, including the United States. The Uruguay Round of GATT negotiations made great strides in extending obligations for fair trade practices to countries worldwide. Many developing countries, however, require technical assistance from industrialized nations in establishing standards systems that comply with their new, often challenging obligations under the Technical Barriers to Trade Agreement of the GATT Uruguay Round. U.S. assistance in this respect will help make the potential gains to U.S. and world trade from the GATT a reality.

Case Example: Emerging Standards and Conformity Assessment Systems in Indonesia

Recent economic reform and institutional change in Indonesia serve as a case example of how evolving national standards systems are important to international trade expansion.[80] New structures, institutions, and policies on standards now being established in emerging economies such as Indonesia will have a significant impact on future U.S. export success. This is particularly true in regard to servicing the dynamic and rapidly growing new markets in the Asia-Pacific region. Assistance by U.S. industry and government in standards and conformity assessment services to countries such as Indonesia can play an important role in enhancing U.S. export success.

Indonesia, along with other East Asian nations, has experienced rapid economic growth over the past decade. GDP in Indonesia grew 5.7 percent over the period 1980-1992. The growth rate of imports in 1992 was 7.1 percent. This economic expansion has been driven by a series of domestic and foreign economic policy reforms launched in 1985. As part of these initiatives, the government liberalized Indonesian trade policies. Reform has included not only a sharp reduction in average tariff rates, but also the elimination of myriad licensing and other quantitative restrictions on imports.[81] There has been substantial progress in deregulation of domestic markets, as well as opening of the Indonesian economy to foreign investment. In part due to these market-opening reforms, U.S. trade with Indonesia has expanded rapidly. In 1993, U.S. exports to Indonesia totaled $2.8 billion. The U.S. stock of foreign investment in 1992 was $4.3 billion.

Economic opening and modernization of the Indonesian manufacturing sector have involved many important policy changes, including a set of changes associated with product standards. Although much of the economy is comprised of government-owned "strategic" corporations, the private economy and nongovernment sector continue to expand. The government, through the National Standardization Council of Indonesia (Dewan Standardisasi Nasional, DSN) has moved to develop a new, modern framework for the development, adoption, and dissemination of product and process standards.[82] There is an opportunity and a need for industrialized countries, such as the United States, to provide important assistance to Indonesia as this work continues. Assistance could be provided both unilaterally, and through regional mechanisms such as the APEC forum.

The DSN and government ministries are currently working to construct new procedures and institutions for (1) product testing and certification, (2) oversight rules for laboratory and certifier accreditation, and (3) rules to accredit the competence of ISO 9000 quality system auditors in Indonesia. The Indonesian government is revising standards and conformity assessment systems to meet the increasing demand for efficiency in the domestic market. These reforms are also motivated by increased competition in foreign markets and the demand for higher quality in Indonesian exports. The type of model Indonesia adopts for an evolving standards system—particularly, the extent to which it centers on either government or private-sector activities—will, therefore, have important implications for future economic growth.

The committees supporting standards activities of the DSN are currently comprised of only 50 percent private-sector representatives. The technical work and drafting of standards are completed by government ministries. There is the expectation that private-sector representation in this work will increase. Expert advice and technical assistance on how to achieve this goal and the evolution of a private standards system are critical. The growth of efficient, private-sector-driven standards development would also serve to benefit foreign firms with manufacturing facilities in Indonesia, as well as exporters wishing to penetrate

Indonesian markets. Through the provision of assistance by in-country experts, training seminars for government officials, and other mechanisms, the U.S. and other governments could play an important role in helping to establish modern and effective standards systems.

There are other ways in which developments in Indonesian standards policies will affect economic development and trade. In order to support global trade expansion and facilitation, it will be necessary to understand and monitor these developments, as they affect U.S. trade and commercial interests. There are currently 3,550 Indonesian standards in force under DSN authority. Only about 20 percent of these are based on international standards. As an outgrowth of trade policy reforms and market liberalization, Indonesia signed the Tokyo Round Standards Code of the GATT in September 1993. Prior to developing a new Indonesian national standard, therefore, the DSN is now obligated to consult ISO and International Electrotechnical Commission (IEC) standards. The DSN has stated a goal of increasing the number of national standards based on international ones, including those developed in the United States. New cement standards in Indonesia, for example, are based on American Society for Testing and Materials standards.

It is important that Indonesia and other new signatories of the Uruguay Round Technical Barriers to Trade Agreement be provided with expert assistance in the implementation of policies to fulfill new obligations under the GATT agreement. This includes aid in creating systems that serve to support multilateral obligations in the agreement, such as increased use of international standards. To the extent that nations such as Indonesia are aided by the United States and other industrialized nations in rapidly meeting new international obligations in standards and conformity assessment, U.S. economic interests are fostered. Countries such as Indonesia will require assistance in meeting the provisions of the new Uruguay Round SPS agreement, for example. This will both support the expansion of Indonesian agricultural exports to industrialized nations, and raise incomes in Indonesia, which serve to expand imports.

In addition to changing policies on the development of national standards, Indonesia is building a new infrastructure to support quality systems management. The manner in which this system develops and the extent to which it is modeled on principles of nondiscrimination, transparency, economic efficiency, and trade facilitation will have a direct impact on both long-term growth in Indonesia, and relations with the United States and other nations. Indonesia adopted the ISO 9000 series as national standards in 1992. As of August 1994, there were 25 private manufacturing firms and one construction company in the service sector with ISO 9000 certification. Another important development centers on the creation of a national accreditation body in 1994 as part of the DSN. In 1994, one quality systems certification body was granted official accreditation, and there were three applications pending as of August 1994. Most of the large testing laboratories in Indonesia are government owned and operated, with smaller, in-house facilities dedicated to servicing private firms. To promote Indonesian export success, the government has emphasized work necessary to gain international recognition of the technical competence of these laboratories. The establishment of advanced systems to support internationally recognized testing facilities is therefore another area in which the expertise and assistance of industrialized nations, including the United States, could be significant, not only to Indonesia but also to the long-term interests of global economic welfare and the facilitation of global trade.

In sum, an active effort by the United States to link standards, trade, and technical assistance is extremely important at this point in history. As developments in nations such as Indonesia indicate, U.S. trading partners in East Asia and Latin America have begun to modernize their standards and conformity assessment infrastructures. U.S. involvement and active participation in assisting these nations to construct policies and programs to meet development needs can serve not only to advance growth prospects overseas, but also help to help bind the United States with these nations as they continue to embrace market-based economic principles and systems.

A Model for Standards Assistance Activities

Since 1990, the U.S. Department of Commerce's National Institute of Standards and Technology has conducted a pilot program to provide technical assistance on standards and conformity assessment to Saudi Arabia. U.S. exports to Saudi Arabia totaled $6.7 billion in 1993. This represents a slight decline from 1992 levels but is significantly higher than U.S. exports of $4.0 billion in 1990. The U.S.-Saudi Arabia Standards Cooperation Program was established to channel technical input from U.S. industry and government experts to the Saudi Arabian Standards Organization (SASO). This program presents a clear model of how an effective standards assistance effort could operate in other U.S. export markets.

The U.S.-Saudi Arabia program centers on the placement of a single, full-time U.S. standards adviser in the capital, Riyadh.[83] The advisers who have filled this position were recruited from private industry. The standards adviser has NIST credentials as a U.S. government representative, raising the position's visibility and authority in the host country. Substantial experience in the private sector with standardization issues is a critical requirement for this position. The standards adviser interacts on a daily basis with SASO officials, providing advice on standards and conformity assessment issues as appropriate. More important, however, is the adviser's role as a communications contact point, channeling requests from SASO for technical information on particular issues back to NIST. These frequently take the form of draft, technical standards prepared by SASO, which U.S. experts are able to review and comment on before they become official Saudi standards.

NIST coordinates referral of questions referred from SASO, including draft standards and certification criteria, to a network of U.S. experts for comment. These experts are not paid. They include standardization officials in private U.S. firms, standards-developing organizations, and government agencies. This referral process is critical to enable a high quality of technical comment on questions and issues raised by SASO. No single expert or group of experts, placed in a host country, could match the breadth and depth of technical expertise accessible through referral of questions back to a network of U.S. firms and agencies. In the model provided by the U.S.-SASO pilot program, the most important role of the standards adviser is to develop communications with host country standards officials and to facilitate their use of the technical assistance mechanism.

The results of the pilot program have been highly promising.[84] From 1990 to 1993, 516 draft SASO standards were sent through this mechanism to the United States for comment.[85] These encompassed standards related to agriculture and food products; construction and building materials; electrical equipment; machinery; chemicals; textiles; and measuring instruments and procedures. U.S. firms with material interests in Saudi standards, primarily U.S. exporters to Saudi Arabia and government agencies, commented on 340 of these drafts. In every case but one, U.S. comments were incorporated directly into formal SASO standards.

Many U.S. comments, for example, have been in the area of automotive industry standards. This sector is the largest single component of U.S. exports to Saudi Arabia, accounting for more than $1 billion in exports in 1993. U.S. automobile manufacturers have been among the most active participants in the U.S.-Saudi pilot program. One key achievement of the program has been to ensure that where SASO standards reference international automotive electronics standards developed by the IEC, they also recognize as equivalent any non-IEC standards used by the U.S. automotive industry. In this way, unnecessary barriers to U.S. exports have been avoided, enhancing the competitive opportunities that the market presents for U.S. manufacturers and preserving the broadest range of choice and competition in the Saudi import market. These benefits were achieved as a result of the rapid, focused access by directly interested U.S. parties made possible by the pilot program.

The most important factors in the success of the U.S.-Saudi Arabia Standards Cooperation Program included the full-time commitment within the host country of a qualified standards adviser; access, through the NIST-coordinated communication channel, to a broad range of U.S. technology and expertise for comment on draft standards; and a high level of participation by U.S. private-sector experts in the draft review and comment process.[86] U.S. industry initially provided matching funds for program expenses. These funds declined over time, and at present, NIST provides full financial support for the program. The most important contribution of U.S. industry to program success, however, continues to be the time and expertise required to prepare written comments and responses to queries referred from the host country standards authorities. In the case of the U.S.-Saudi pilot program, the level of U.S. private participation in this respect has not declined.

As of October 1994, the program was preparing to provide a second type of assistance to Saudi Arabia that also has strong potential applicability in emerging U.S. export markets. In the fall of 1994, a delegation from the American Automobile Manufacturers Association was scheduled to conduct a technical seminar for SASO's engineering staff to update them on a range of state-of-the-art automobile safety technologies. The seminar was to present information on items such as air bags; conversion of automobiles to unleaded gasoline; antilock brakes; electronic suspension; passive seat restraints; and design for distribution of impact energy in a collision. Conveying technical information on these topics to the Saudi officials responsible for writing product certification requirements has at least two crucial benefits: (1) it improves SASO's capacity to write effective safety standards for Saudi Arabia; and (2) it promotes the use of U.S. industrial technologies in an important market for U.S. manufacturers.

Duplication of the U.S.-Saudi Arabia export promotion program in other foreign markets will depend on strong linkages and communication between the U.S. government and private industry. It is particularly important that programs of the Department of Commerce, NIST, and other agencies involved in the promotion of U.S. exports are coordinated. In the Department of Commerce Appropriations Act for 1995, the Congress appropriated approximately $5 million for an expanded program within NIST on International Trade Standardization and Measurement Services.[87] A review of NIST plans for expansion of assistance in this program, which builds on the SASO model, reveals the following conclusions:

1. An expanded program in standards assistance overseas should involve U.S. private sector firms closely both in providing advice to host country officials, and in organizing special government–industry training missions to developing and newly industrializing nations, especially those in East Asia and Latin America. These special missions should be funded by industry.

2. This program should target a wide number of developing countries in East Asia and Latin America, where U.S. standards assistance will have the most significant impact on the evolution of standards and conformity assessment systems. Current plans call for the placement of experts in Russia, the Czech Republic, Mexico, India, China, Japan, Korea, Argentina (for all South America), Geneva (for the United Nations Economic Commission for Europe), and Paris (for the OECD nations). Although it is anticipated that these missions will serve multiple markets, a more rapid expansion of personnel in individual countries is necessary to take full advantage of rapidly developing systems overseas.

3. An overseas assistance program should provide regular, ongoing written and other reports on standards and conformity assessment developments to the Department of Commerce's trade promotion programs, Office of the U.S. Trade

Representative, Export-Import Bank, and other relevant U.S. government agencies.

SUMMARY AND CONCLUSIONS

This chapter has outlined both the challenges and the opportunities offered by standards and conformity assessment systems overseas. They represent both the potentially complex and difficult barriers to international trade and unique opportunities through innovative export promotion programs to increase U.S. competitiveness in global markets. Future U.S. policy in these areas, therefore, will play an important role in positioning U.S. industry for success into the next century.

Several key developments in the international economy will serve to heighten the importance of standards and conformity assessment as barriers to future global trade. These include (1) the decline of tariff barriers in world markets, especially as the Uruguay Round of tariff cuts is implemented; (2) the growing complexity of conformity assessment mechanisms in both the industrializing and the developed nations; and (3) new demands for third-party assurance, not only of product safety, but also of producers' quality management systems and environmental management systems, among other factors.

There are several necessary conditions for the United States to take advantage of the growing interrelationships among standards, conformity assessment, and global trade. First, a high level of attention to emerging standards developments by both government and private industry is necessary. In particular, there is the need for a new analytical capacity in government to monitor and report on the economics of standards and conformity assessment, including domestic systems, as well as developments in overseas markets. The specific functions of such a unit are outlined in greater detail in the following chapter.

Second, strong support for mechanisms and programs by the United States to rapidly implement the provisions that support trade liberalization in the Uruguay Round Agreement is necessary. This should include the type of organized and detailed support for implementation that is being conducted by the United States through the APEC standards and conformity, for example. Moreover, the U.S. government, in consultation with industry, must continue to provide leadership on the conclusion of MRAs with the European Union and other major trading partners, particularly in the APEC region. This will help ensure continued progress in the world economy in reducing barriers to trade in standards, especially under the likely circumstances that a major trade negotiating round will not be under way for a long period of time.

In addition to U.S. leadership in negotiations on bilateral or regional MRAs and the provision of technical assistance, a post-Uruguay Round U.S. trade policy should involve aggressive use of the new dispute resolution procedures available in the WTO to address barriers related to standards and conformity assessment.

U.S. policy must also leverage our ability to move in a unilateral fashion to open overseas markets. This should come through a more proactive use of Section 301 to remove foreign barriers to U.S. exports. Finally, U.S. trade policy should work toward rapid development of mutual recognition agreements with major trading partners in order to facilitate trade expansion and forestall the development of new barriers in the post-Uruguay Round system. These policy prescriptions are developed more fully in Chapter 5, along with detailed recommendations for government and industry, as requested by Congress in the solicitation of this report.

NOTES

1. David Walters, *Comparison of Export Sector Wages to Overall Sector Wages.* Additional data supplied by the Office of the Assistant U.S. Trade Representative for Economic Affairs, Office of the U.S. Trade Representative, Washington, D.C.

2. Data supplied by the Office of the Assistant U.S. Trade Representative for Economic Affairs, Office of the U.S. Trade Representative.

3. John Wilson, *The U.S. Performance in Advanced Technology Trade: 1982-93 (est.).*

4. Office of the Chief Economist, Office of the U.S. Trade Representative, *U.S. Exports Create High-Wage Employment.*

5. Kindleberger, *Standards as Public, Collective and Private Goods*, 384-385.

6. As noted in previous chapters, there is a growing body of economics research on compatibility standards. See, for example, Besen, Stanley M., and Leland L. Johnson. 1986. *Compatibility Standards, Competition, and Innovation in the Broadcasting Industry.*

7. There are numerous studies which detail trade protection in the post-war period. Among the recent analyses of the types of protection employed by the industrialized and developing nations in the post-war period, especially those under discussion in the Uruguay Round of multilateral trade negotiations see, for example; *Analytical and Negotiating Issues in the Global Trading System*, A.V. Deardorff and R. M. Stern. *The Uruguay Round: An Assessment*, Jeffrey J. Schott. *American Trade Politics*, 2nd edition, I. M. Destler, among many others.

8. Estimates of the cost of tariff and non-tariff trade barriers have been conducted by numerous organizations. See, for example, recent work of multilateral institutions such as the Organization for Economic Cooperation and Development (OECD), Paris, and World Bank, as well as the Center for the Study of American Business, University of Washington, St. Louis, and U.S. government agencies, such as the United States International Trade Commission (USITC).

9. See, for example, Jaime de Melo and David Tarr, *A General Equilibrium Analysis of U.S. Foreign Trade Policy.*

10. White House, *Trade Agreements Resulting from the Uruguay Round of Multilateral Trade Negotiations*, Federal Register 58, no. 242: pp. 67269-67270.

11. See, for example; Sam Laird and Alexander Yeats, *Quantitative Methods for Trade Barrier Analysis.* United Nations Council on Trade and Development (UNCTD), *Problems of Protectionism and Structural Adjustment: Restrictions on Trade.* World Bank, *World Development Report 1987*, among other reports.

12. Joseph Grieco, *Cooperation Among Nations: Europe, America, and Non-Tariff Barriers to Trade.*

13. For a discussion, see the studies summarized in *Trading Free, The GATT and U.S. Trade Policy*, Patrick Low, 73-74.

14. *World Development Report 1987*, The World Bank.

15. As part of this study, extensive interviews were conducted with senior U.S. government

officials responsible for trade policy formation and implementation. These included briefings by officials from USTR and Department of Commerce concerning barriers to U.S. exports related to standards and conformity assessment policies overseas. Information was also gathered in Tokyo, Japan and Jakarta, Indonesia in August 1994 from (1) U.S. government officials in the U.S. and Foreign Commercial Service, economic and commercial affairs, and science and environment offices, and (2) U.S. business executives located in Japan and Indonesia through the American Chamber of Commerce.

16. Office of the United States Trade Representative, *1994 National Trade Estimates Report on Foreign Trade Barriers.*

17. *1994 National Trade Estimate Report of Foreign Trade Barriers*, 93-94.

18. U.S. Government Trade Policy Coordinating Committee, Product Standards Working Group, estimates applied to U.S. Census Bureau export data, Washington, 1994.

19. *1994 National Trade Estimate Report of Foreign Trade Barriers*, 81.

20. Data provided by Charles Ludolph, Director, Office of EC Affairs, U.S. Department of Commerce, October 27, 1993.

21. Data supplied by Charles Ludolph, Director, Office of European Affairs, International Trade Administration, U.S. Department of Commerce, 1994.

22. This section draws, in part, on Victoria Curzon Price, *The Post Uruguay Round Trade Outlook: Standards and Technical Barriers to Trade,* and John Sullivan Wilson, S*tandards, Conformity Assessment, and Trade: New Developments and the Asia Pacific Economic Cooperation (APEC) Forum.*

23. Office of the United States Trade Representative, *Final Texts of the GATT Uruguay Round Agreements*, including the agreement establishing The World Trade Organization, and *Agreement on Technical Barriers to Trade*, 117-137.

24. Office of the United States Trade Representative, *Final Texts of the GATT Uruguay Round Agreements*, Agreement of the Application of Sanitary and Phytosanitary Measures, p. 69-83. For an overview of the Agreement and implications for the U.S. see: Donna U. Vogt, *Sanitary and Phyto-sanitary Safety Standards for Foods in the GATT Uruguay Round Accords.*

25. For a comprehensive assessment of the implications of the Uruguay Round Agreements in full see: Schott, *The Uruguay Round: An Assessment.*

26. For an analysis of the 1994 Agreement see Richard H. Steinberg, *The Uruguay Round: A Legal Analysis of the Final Act*, 1-97.

27. Calculated using World Bank, IMF data on imports for 1991, totals for signatories to the Tokyo Round Standards Code and new signatories to the Uruguay Round Agreement.

28. Data included in William J. Clinton, *Report to Congress on Recommendations on Future Free Trade Area Negotiation*, Table 2.

29. Based on estimates by NRC Project on International Standards, Conformity Assessment, and U.S. Trade Policy. Calculated from data supplied by the U.S. Bureau of the Census, 1993-1994.

30. Locke, John W. 1993. *Conformity Assessment—At What Level?*

31. Curzon-Price, *The Post-Uruguay Round Trade Outlook: Standards and Technical Barriers to Trade.*

32. For an overview of efforts to develop environmental management system standards at the international level see: Marilyn R. Block, *ISO/TC 207: Developing an International Environmental Management Standard.*

33. Harmonization of technical regulations as well as certification mechanisms have been particularly significant in health-related industries, such as medical devices. See, for example, Health Industry Manufacturers Association, *EU-U.S. Mutual Recognition Agreements (MRAs): Key Issues for the Medical Device Industry.*

34. Office of Public Affairs, Office of the United States Trade Representative. Facsimile describing the USTR and its functions.

35. U.S. Congress, House, Committee on Ways and Means, *Overview and Compilation of U.S. Trade Statutes*, 1993 Edition, p. 185-196.

36. Office of Public Affairs, Office of the United States Trade Representative. Facsimile describing the USTR and its functions.

37. Subtitle A of Title VII, Section 704 of the *Tariff Act of 1930* deals with termination or suspension of investigations involving countervailing duties. Subtitle B of Title VII Section 734 deals with the termination or suspension of investigations involving antidumping duties.

38. Title III, Chapter 1, Section 301 of the *Trade Act of 1974 (amended)* outlines the actions that may be taken by the USTR to enforce U.S. rights under the trade agreement and as a response to certain foreign trade practices.

39. U.S. Congress, House. Committee on Ways and Means, *Overview and Compilation of U.S. Trade Statutes*, 1993 Edition, p. 185-196.

40. Ibid.

41. In *1993 Annual Report*, Office of the United States Trade Representative.

42. Office of Intergovernmental Affairs, Office of the United States Trade Representative. Facsimile describing the advisory committee system and its functions.

43. Office of United States Trade Representative, *1993 Annual Report*. Facsimile transmission of pp. 114-117.

44. Ibid.

45. For an overview of Section 301, including USTR self-initiation of cases under the statute see: John S. Wilson, *The U.S. Government Trade Policy Response to Japanese Competition in Semiconductors: 1982-87.*

46. John H. Jackson and William J. Davey, *Legal Problems of International Economic Relations*, p. 804.

47. Patrick Low, *Trading Free: The GATT and US Trade Policy*, p. 94.

48. John H. Jackson and William J. Davey, *Legal Problems of International Economic Relations*, p. 147.

49. Patrick Low, *Trading Free: The GATT and US Trade Policy*, p. 95.

50. Data in this section is drawn from Patrick Low, *Trading Free: The GATT and US Trade Policy*, and from Thomas O. Bayard and Kimberly Ann Elliott, *Reciprocity and Retaliation in U.S. Trade Policy.*

51. The results of this work are cited in a recent analysis of the use of Section 301; *Reciprocity and Retaliation in U.S. Trade Policy*, Thomas O. Bayard and Kimberly Ann Elliott. For further detail see also, Gary Clyde Hufbauer, Jeffery J. Schott, and Kimberly Ann Elliott, *Economic Sanctions Reconsidered.*

52. Industry and government assessment of the negotiations and outcome are outlined in *Reciprocity and Retaliation in U.S. Trade Policy*, Thomas O. Bayard and Kimberly Ann Elliott, p. 134-138.

53. For an overview of the NAFTA agreement, including reference to standards and conformity assessment issues in NAFTA see; U.S. International Trade Commission, *Potential Impact on the U.S. Economy and Selected Industries of the North American Free Trade Agreement.* Gary Clyde Hufbauer and Jeffrey J. Schott, *NAFTA: An Assessment*, among others.

54. Clinton, *Report to the Congress on Recommendations on Future Free Trade Area Negotiations.*

55. New European systems also exist for facilitating trade in non-regulated products. The European Organization for Testing and Certification (EOTC) was established in Brussels in 1992. EOTC functions as a focal point for promoting mutual confidence among private-sector conformity assessment systems. It recognizes voluntary "agreement groups" consisting of testing, certification, and accreditation organizations in a range of industry sectors. EOTC has approved participation of non-European organizations in the agreement groups, indicating that the new system is unlikely to present a barrier to U.S. exports. The organization is new, however, and continued monitoring of its

impact on international trade is necessary before conclusions can be drawn. See EOTC, *EOTC: Focal Point for Testing & Certification in Europe.*

56.   For a detailed description of European Union harmonized standards and certification procedures for regulated products, see U.S. Department of Commerce, *EC Product Standards Under the Internal Market Program* and U.S. Department of Commerce, *EC Testing and Certification Procedures Under the Internal Market Program.*

57.   For example, U.S. information technology firms were highly concerned in 1994 about a proposed ETSI intellectual property policy. This would have required firms to license their proprietary technology for inclusion in ETSI standards at unfavorable and inflexible terms. Opposition from U.S. firms and government officials persuaded ETSI to revise the policy. Examples of successfully resolved disagreements with CEN are presented in an internal memorandum from ANSI's Brussels office to Manuel Peralta, then ANSI president. The memorandum states in part, "...the CEN responses to the U.S. concerns have been positive and are almost complete. The system is working well in the standardization area. The only pending case studies requiring a response from CEN concern power lawn mowers and toy safety." D. W. Smith, Vice President, ANSI Brussels Office, memorandum, January 8, 1993.

58.   For a thorough discussion of product regulation and conformity in the EU (formerly, the European Community, or EC), see U.S. Department of Commerce, International Trade Administration, *EC Testing and Certification Procedures Under the Internal Market Program.*

59.   For a discussion of the early results of these talks see: Charles M. Ludolph, *Mutual Recognition Agreements—Access to the European Union.*

60.   Charles M. Ludolph, *Mutual Recognition Agreements, Part III: Summary of Negotiations at Midpoint.*

61.   Richard Schulte, Vice President, AGA Laboratories, personal communication, June 30, 1994.

62.   See Ludolph, *Mutual Recognition Agreements, Part II.* For a discussion of specific issues in electromagnetic interference testing, see Bert G. Simson, *Conformity Assessment Workshop on Electromagnetic Compatibility,* and Retlif Testing Laboratories, *U.S./EU Trade Negotiations: EMC Issues.*

63.   For discussion of specific issues arising in the U.S.-EU negotiations, see Ludolph, *Mutual Recognition Agreements—Access to the European Union* and Ludolph, *Mutual Recognition Agreements Part II.*

64.   See, for example, Donald S. C, Deputy U.S. Trade Representative, quoted in Grieco, *Cooperation Among Nations: Europe, America, and Non-Tariff Barriers to Trade,* p. 194.

65.   National Institute of Standards and Technology, *Establishment of the National Voluntary Conformity Assessment System Evaluation Program,* p. 19129-19133.

66.   Ludolph, *Mutual Recognition Agreements—Access to the European Union.*

67.   ANSI, letter to Charles F. Meissner, Assistant Secretary for International Economic Policy, U.S. Department of Commerce, and Richard G. Meier, Deputy Assistant U.S. Trade Representative, April 20, 1994.

68.   It is important to note that even with the CE Mark, the free flow of goods throughout the European Union has not yet been fully realized. The acceptance of testing and certification performed in the less-developed member states by national authorities elsewhere in Europe is mandated by the European Commission, but does not always, in practice, occur. See Ludolph, *Mutual Recognition Agreements*; and David Stanger, *EOTC.*

69.   For discussion of product liability in the U.S., including sector-specific case studies, see National Academy of Engineering, *Product Liability and Innovation.* See also Eads and Reuter, *Designing Safer Products.*

70.   See, for example, *Report on United States Barriers to Trade and Investment, 1994,* Services of the European Commission, Section G, "Standards, Testing, Labelling, and Certification, p. 55-60.

71. *NVLAP, National Voluntary Laboratory Accreditation Program, 1994 Directory*, U.S. Department of Commerce, Technology Administration, National Institute of Standards and Technology, and American Association for Laboratory Accreditation, *A2LA 1994 Directory of Accredited Laboratories*.

72. Clinton, *Report to the Congress on Recommendations on Future Free Trade Area Negotiations*.

73. Ibid.

74. For an overview of possible directions for the APEC forum, including issues of trade facilitation, see Gary Clyde Hufbauer, *Whither APEC?*

75. *Achieving the APEC Vision; Free and Open Trade in the Asia Pacific*, Asia-Pacific Economic Cooperation.

76. *APEC Economic Leader's Declaration of Common Resolve*, Bogor, Indonesia, November 15, 1994.

77. *Proposed Declaration on an APEC Standards and Conformance Framework: Submitted by Australia*, APEC Committee on Trade and Investment.

78. U.S. International Trade Commission, *Development Assistance in East Asia*, p. 98-99.

79. Ibid.

80. This section draws, in part, on information presented in interviews conducted with U.S. government officials, affiliates of the U.S. Chamber of Commerce in Jakarta, Indonesia, and representatives of the Indonesian government, including the Dewan Standardisasi Nasional (DSN), Standardization Council of Indonesia, in Jakarta, Indonesia, in August 1994. Personal communications with John Wilson, project director.

81. For an overview of recent trade reforms in East Asia, including Indonesia, see: *Development in Practice, East Asia's Trade and Investment, Regional and Global Gains from Liberalization*, The International Bank for Reconstruction and Development.

82. DSN operates under authority of a Presidential Decree of 1984, revised in 1989 to establish Indonesian national standards (Standard Nasional Indonesia, SNI). DSN is the coordinating body through which all standards and metrology organizations establish standards in Indonesia. It works with 2,000 experts from industry and government.

The DSN represents Indonesia as the government members body of the International Organization for Standardization (ISO), the International Electrotechnical Commission (IEC) and represents Indonesia in Codex Alimentarius Commission (CAC). Pusat Standardisasi LIPI operates as the Secretariat of the DSN. It manages Indonesian participation in the activities of ISO and IEC. Technical committees employ experts from government, private institutions, industry, and professional associations.

83. Information on the U.S.-Saudi Standards Cooperation Program provided by Gilbert Dwyer, American-Saudi Roundtable, and Edward Wunder, NIST representative to SASO, fact sheets and personal communication, May 16, 1994.

84. For example, an independent advisory panel of industry and academic experts reviewed this program in 1993 and recommended that NIST's proposal to expand it be fully funded. See Visiting Committee on Advanced Technology of the National Institute of Standards and Technology, *International Standards Issues: A Statement to the Secretary of Commerce*, p. 4-6.

85. American-Saudi Roundtable, *U.S.-Saudi Arabia Standards Cooperation Program*, progress report dated December 31, 1993.

86. Memorandum by Gilbert Dwyer, American-Saudi Roundtable, *U.S.-Saudi Arabia Standards Cooperation Program: Critical Success Factors*, March 21, 1994. For an additional example of private standards setting firms assisting in other countries see: *Letter of Intent between NSF and The Instituto Mexicano de Tecnologia del Agua*, in which the NSF agreed to share through training its expertise and experience in product certification with the ultimate goal of mutual recognition.

87. U.S. Congress. House. *Department of Commerce Appropriations Act, 1995*.

5

Recommendations to Address Future Challenges and Opportunities

This report has examined domestic and international processes for developing standards and assessing conformity to them, as well as the relationships between these processes and U.S. export success. It has demonstrated that standards and conformity assessment systems have important influences on international trade. These are manifest in the successes and opportunities for trade expansion, such as strengthened disciplines in the Uruguay Round Technical Barriers to Trade Agreement provisions; various non-tariff trade barriers that must be reduced or eliminated; and underutilized tools for export expansion. The efficient functioning of standards and conformity assessment systems represents a significant future challenge to industrial advance in the United States, as well as an opportunity for leveraging national advantages in global markets.

This chapter summarizes the state of the U.S. standards and conformity assessment systems and outlines recommendations to enhance their operation to promote economic efficiency. The recommendations include suggested changes in public policies to improve the domestic system, as well as to position U.S. trade policy initiatives to support international trade and U.S. economic advance. As noted below, some recommendations call for action by Congress in the form of new legislation, while others can be implemented by executive branch agencies under existing legislative authority.

The state of the U.S. conformity assessment system poses the most immediate and direct challenge to economic success. Recommendations to streamline this system and increase its efficiency are contained in this chapter. The U.S. standards development system, by comparison, functions well and is effective at meeting national interests. There is the need for improvement, however, in public–private cooperation in standards development and government use of private standards. Recommendations on international trade policy in this chapter focus on proactive efforts to enhance present and future export opportunities. Finally, recommendations are made to strengthen the nation's capacity to acquire, analyze, and disseminate critical information about international standards and conformity assessment. Measures to anticipate and deter future barriers to trade linked to standards and certification are also outlined.

CONFORMITY ASSESSMENT

Previous studies of the U.S. standards system have emphasized the processes by which standards are developed.[1] As this report has demonstrated, however, these mechanisms account for only part of the economic and societal impact of standards. The increasingly complex U.S. and international mechanisms for assessing product and process *conformity* to standards are also significant. These mechanisms include product testing and certification; certification of manufacturing processes such as quality control systems; accreditation of laboratories and certifiers; and government recognition of accreditors, among others. As Chapter 3 outlines, there is increasing demand by customers and government regulators for independent (third-party) assurance of conformity, both in the United States and abroad. The growing complexity of the system imposes uncertainty and cost on U.S. manufacturers. It poses challenges to public and private actors in the U.S. conformity assessment system to keep pace with rapid change.

Significant improvement is needed in the U.S. system for assessing conformity of products and processes to standards. Our system has become increasingly complex, costly, and burdensome to national welfare. This is reflected in unnecessary duplication and unwarranted layers of complexity at the federal, state, and local levels. Manufacturers are increasingly forced to perform redundant tests and obtain repetitive certifications for products sold in different parts of the country. Testing laboratories pay unnecessary fees and undergo duplicative audits to demonstrate their competence to multiple federal, state, and local authorities. The result is higher costs for U.S. manufacturers, public procurement agencies, testing laboratories, product certifiers, and consumers.

Data on the precise magnitude of these costs in the U.S. economy are lacking. The rapid growth of U.S. independent testing services, currently accounting for more than $10 billion in annual revenue, is nevertheless an indication of the expansion of the conformity assessment system. Chapter 3 contains many specific examples of duplications in product testing and accreditation. In addition, a 1993 study of environmental testing laboratory accreditation commissioned by the Environmental Protection Agency (EPA) identified a potential for nationwide cost reduction of approximately 28 percent.[2] This saving could be achieved through elimination of redundant accreditation requirements among local, state, and federal agencies.

Although no similar study has been performed in the area of product testing, anecdotal evidence, consultation with a wide range of experts, and a review of federal and state accreditation programs support the conclusion that a similar level of inefficiency exists in that area. A precise estimate of the cost of inefficiency in U.S. state and local systems is outside the scope of this study. Available evidence and analysis, however, demonstrate the need for action to streamline the system of product testing.

At the federal level, government agencies should retain responsibility for oversight of critical regulatory and procurement standards in areas of preeminent public health, safety, environmental, and national security concerns. The assessment of product conformity to those standards, however, is essentially a technical function performed most efficiently and effectively by the private sector. Government should meet its responsibility for serving the public interest in an oversight capacity.

The federal role in conformity assessment should center on *recognition* of private-sector services. Government should evaluate and recognize private sector organizations that are competent to accredit testing laboratories, product certifiers, and quality system registrars.[3] In order to streamline and improve national conformity assessment procedures, the following is recommended:

- **RECOMMENDATION 1:** Congress should provide the National Institute of Standards and Technology (NIST) with a statutory mandate to implement a government-wide policy of phasing out federally operated conformity assessment activities.

    NIST should develop and implement a National Conformity Assessment System Recognition (NCASR) program. This program should recognize accreditors of (a) testing laboratories, (b) product certifiers, and (c) quality system registrars. By the year 2000, the government should rely on private-sector conformity assessment services recognized as competent by NIST.

A properly implemented NCASR program will reduce costs for federal agencies by eliminating the need to design and operate government-unique and duplicative testing, certification, and accreditation programs. It will decrease costs for industry of complying with separate, duplicative private and public conformity assessment requirements. The program will draw on NIST's expertise in testing and evaluation, as well as its in-house scientific and technical resources. It will also incorporate, under NIST guidance, participation of technical experts from regulatory and procurement agencies. These agencies will continue to be responsible for determining guidelines for essential standards of safety, health, environmental protection, and fitness for public procurement against which products are assessed. A congressional mandate for NIST should state explicitly, however, that reliance on NIST-recognized accreditation programs is the means by which agencies should determine compliance with product regulations and procurement standards under their jurisdiction. The deadline of the year 2000 for a phaseout of direct government-operated conformity assessment activities should allow the necessary time for NIST to provide assurances of private-sector entities to take on these responsibilities. This is, in fact, already under way in several areas.

Two current programs serve as precedents for the recognition function recommended here. The first is NIST's responsibility under the Fastener Quality Act (P.L. 101-592) to ensure the competence of laboratories that test fasteners. NIST is charged by the act not only with accrediting testing laboratories directly, but also with evaluating and *recognizing* accreditors.[4] A second program that might serve, in part, as an example for NCASR is the National Voluntary Conformity Assessment System Evaluation (NVCASE) program, established by NIST in 1994. NVCASE is designed to evaluate and formally recognize U.S. accreditors of laboratories, certifiers, and quality system registrars in a broad range of product sectors. NVCASE is a voluntary program, and it is applicable only to assessment of U.S. exported goods against foreign regulations. The NCASR program, in contrast, should apply to all conformity assessment activities for compliance with *domestic* regulatory and procurement standards.

Inefficiency in the U.S. conformity assessment system is especially apparent at the state and local levels. There have been only extremely limited national efforts, however, to promote mutual acceptance by state and local authorities of product testing results, certifications, and laboratory accreditations. In order to reduce redundancy and inefficiency at the state and local levels of the U.S. conformity assessment system, the following is recommended:

- **RECOMMENDATION 2:** NIST should develop, within one year, a ten-year strategic plan to eliminate duplication in state and local criteria for accrediting testing laboratories and product certifiers. NIST should lead efforts to build a network of mutual recognition agreements among federal, state, and local authorities.

  After 10 years, the Secretary of Commerce should work with federal regulatory agencies to eliminate remaining duplication through preemption of state and local conformity assessment regulation.

Eliminating duplication in state and local conformity assessment regulations does not involve elimination of all differences among federal, state, and local regulations. There are valid reasons for variations in some standards. It is appropriate, for example, that building codes are more stringent in California for earthquake protection and in Florida for resistance to hurricane-force winds. Competently performed testing and certification, however, should be accepted by all state and local authorities throughout the United States.

The details of a strategic plan to eliminate duplication should be developed by NIST, in consultation with other federal, state, and local agencies and the private sector. The NIST-led National Conference on Weights and Measures (NCWM) is a promising model for such a plan. The NCWM, as outlined in Chapter 3, promotes regular consultation among state and federal agencies on technical procedures for weighing and measuring products shipped within the United States. The federal–state partnership fostered under this structure has served the needs of domestic commerce. Cooperation in the broader area of product conformity has the potential to yield benefits on a similar scale.

Clearly, political resistance to change is likely to be strong in some states and localities. For this reason, motivation for progress must be ensured through a congressionally mandated 10-year deadline for agreement. After this period, federal conformity assessment regulations should preempt remaining duplications. A clear precedent for this type of preemption exists in the Occupational Safety and Health Administration's (OSHA's) Nationally Recognized Testing Laboratory program. State and local agencies are prohibited, through preemption by Department of Labor regulations, from refusing to accept product testing data and certification performed by OSHA-recognized laboratories.[5]

STANDARDS DEVELOPMENT

Chapters 1 and 2 of this report examine the role of standards in a modern economy and assess the processes by which standards are developed. *The U.S. standards development system serves the national interest well.* In most cases, it supports efficient and timely development of product and process standards that meet economic and public interests. The system is complex and diversified, with a lead role for the private sector in decisionmaking. The system's decentralized structure provides for flexibility in meeting new market and societal needs. It embodies, therefore, the mechanisms necessary to respond to technological change and the uncertainties of market-driven economic advance.

This report has identified a serious need for improvement, however, in federal policies and public–private cooperation to increase federal use of voluntary consensus-based standards developed in the private sector. There are many benefits to government use of standards developed in the private sector for regulatory and public procurement needs.[6] Government adoption of private standards lowers costs to federal agencies and taxpayers. It also reduces unjustifiable burdens on private firms to meet duplicative standards for both government and private markets. Clearly, not every public standard can be developed through private-sector processes. Government should, however, rely on private activities in all but the most vital cases involving protection of public health, safety, environment, and national security. The U.S. government should establish improved mechanisms to ensure progress toward reaching this goal.

Current efforts by the U.S. government to leverage the strengths of the U.S. standards establishment and its services are inadequate. There has been only limited progress by federal agencies in promoting the use of private standards. The Office of Management and Budget (OMB) Circular A-119, "Federal Participation in the Development and Use of Voluntary Standards," does set a sound goal for increased federal adoption of private, consensus standards. As discussed in Chapter 2, however, the circular provides an ineffective mechanism to ensure government action. Stronger, better-coordinated institutional mechanisms are needed. These should be led by NIST, which has unique technical expertise to coordinate government interaction with private standards organizations, as well as federal use of private standards.

A legislative mandate for NIST oversight of federal adoption of consensus standards will accomplish these critical tasks: It will (a) significantly improve and streamline the government's regulatory and procurement functions; (b) reduce burdens on U.S. manufacturers arising from duplicative government and commercial standards; and (c) provide a centralized government focus for liaison with private sector standards organizations, including—but not limited to—the American National Standards Institute. The following is therefore recommended:

- **RECOMMENDATION 3:** Congress should enact legislation replacing OMB Circular A-119 with a statutory mandate for NIST as the lead U.S. agency for ensuring federal use of standards developed by private, consensus organizations to meet regulatory and procurement needs.

The existing requirement for reports to OMB on agency use of private standards has not proved sufficient to ensure progress toward the policy goals expressed in Circular A-119. A congressional mandate that NIST report annually to Congress on progress within all agencies in using private standards would significantly increase the effectiveness of these policies. NIST work to fulfill this mandate in the interagency process will not be without difficulty; however, over time the leadership exercised by NIST should provide for significant progress. In addition, the Interagency Committee for Standards Policy (ICSP) should be reconstituted. The ICSP's present structure, comprising 35 members in equal standing from all federal agencies with major or minor standards activities, is too large and inflexible. A stronger leadership role for NIST and active participation by core standards-using agencies such as the Department of Defense, the General Services Administration, EPA, OSHA, and the Consumer Product Safety Commission would enable the ICSP to coordinate interagency decisionmaking more effectively. Other agencies should participate in information sharing and decisionmaking on issues relevant to their respective missions.

In addition to intragovernment coordination, there is a need for effective, long-term public–private cooperation in standards development and use. As Chapter 2 outlines, this type of cooperation has often been lacking, although improvement has occurred in recent years. Sustained cooperation requires improved information transfer between government and industry; clear division of public and private responsibilities within the consensus standards system; and institutionalized mechanisms to effect systems improvement and lasting change. The following is therefore recommended:

- **RECOMMENDATION 4:** The Director of NIST should initiate formal negotiations toward a memorandum of understanding (MOU) between NIST and the American National Standards Institute (ANSI). The MOU should outline modes of cooperation and division of responsibility between (1) ANSI, as the organizer and accreditor of the U.S. voluntary consensus standards system and the U.S. representative to international, non-treaty standards-setting organizations; and (2) NIST, as the coordinator of federal use of consensus standards and recognizing authority for federal use of private conformity assessment services. NIST should not be precluded from negotiating MOUs with other national standards organizations.

  In addition, all federal regulatory and procurement agencies should become dues-paying members of ANSI. Dues will support government's fair share of ANSI's infrastructure expenses.

An ANSI-NIST MOU would not be exclusive to these two parties. NIST could sign MOUs with other standards developers, for example, to facilitate use of their standards as appropriate to meet government needs. ANSI could enter into agreements with other agencies. The purpose of an ANSI-NIST MOU is to institutionalize a public division of the unique responsibilities of ANSI and NIST within the U.S. and international standards systems. The MOU will also raise the visibility, within and outside government, of federal policies mandating use of private consensus standards.

Government payment of ANSI dues, similarly, would not preclude use of non-ANSI standards or membership in standards organizations aside from ANSI. Dues payment would, however, uphold the federal government's responsibility, as a major beneficiary of the private, ANSI-accredited consensus standards system, to support its share of ANSI infrastructure and expenses. Dues from federal, state, and local government agencies presently account for less than 1 percent of ANSI dues collections and less than half of 1 percent of total ANSI revenues. Government adoption of private standards developed under the ANSI-accredited system, however, as well as participation of government experts in standards-writing committees, imposes substantial administrative costs on ANSI. These costs include managerial oversight of standardization processes; information dissemination and communication; and support for conferences and technical committee meetings, among others. Government payment of ANSI dues will enhance ANSI's ability to meet these administrative burdens and will provide for improvements in communication between the public and private standards communities. Agency dues will not, however, represent so large a share of ANSI's revenues as to compromise its capacity for independent oversight of consensus standards processes. Dues will be limited to a level sufficient to meet administrative burdens. They will not provide, and should not be considered, a broad public subsidy of the private standards development system.

INTERNATIONAL TRADE

Expansion of U.S. exports is a vital economic interest. As Chapter 4 has outlined, exports promote a strong domestic economy, increased productivity, and employment opportunities for U.S. workers. Moreover, although the United States is the most open economy in the world, continued work to promote a domestic economy open to foreign trade and competition is also important. Two integral components of U.S. trade policy aimed at securing the benefits of trade are (1) reduction of trade barriers, and (2) promotion of U.S. exports. U.S. trade policy, with input from U.S. industry, is targeted at accomplishing these objectives through multilateral, regional, and bilateral efforts. The U.S. Trade Representative (USTR), in cooperation with the Department of Commerce's International Trade Administration, should make every effort to encourage our trading partners to adopt transparent and open standards and conformity assessment systems.

At the multilateral level, the Uruguay Round negotiations of the General Agreement on Tariffs and Trade (GATT) achieved significant progress in expanding global trade and reducing trade barriers—particularly those associated with discriminatory national standards and product certification systems. *The growing complexity of standards and conformity assessment systems in many nations, however, threatens to undermine future trade expansion.* Manufacturers throughout the world face increasing costs in gaining product acceptance in multiple export markets. Many nations impose unnecessarily duplicative or discriminatory requirements for product testing, certification, and quality assurance.

The European Union's (EU's) mechanisms for approving regulated products, in particular, are a barrier to U.S. exports. The EU requires that product certification be performed by European testing laboratories and certifiers designated by European national governments. The system imposes unbalanced costs on U.S. manufacturers as they seek to compete in European markets, for example, by requiring retesting of products already certified within the United States. It also prevents U.S. testing laboratories from providing service to U.S. manufacturers for export of EU-regulated products, except in the limited capacity of subcontractors to European laboratories. As noted in Chapter 4, the severity of these obstacles varies by industry sector. It is important, however, to achieve a rapid, negotiated removal of EU barriers to U.S. exports, both to expand trade with Europe and to set a model for similar negotiations between the United States and other trading partners.

U.S.-EU negotiations toward mutual recognition of conformity assessment procedures began in 1994. *The objective of mutual recognition agreements (MRAs) is to enable manufacturers to test products once and obtain certification and acceptance in all national markets.* The negotiations are led on the U.S. side by the USTR, with input from the Department of Commerce, other federal agencies, and private U.S. firms. Negotiators are seeking MRAs in 11 EU-regulated sectors: information technology; telecommunications equipment; medical devices; electronics; electromagnetic interference; pharmaceuticals; pressure equipment; road safety equipment; recreational boats; lawn mowers; and personal protective equipment such as helmets. These MRAs have strong potential to overcome impediments to trade and increase U.S. manufacturers' access to export markets.

Early discussions within the Asia Pacific Economic Cooperation (APEC) forum in 1994 indicate a favorable economic and political climate for regional MRAs in the Asia-Pacific region. Members of the APEC forum, including the United States, account for 40 percent of global economic activity. Many of the world's fastest-growing economies belong to APEC. Trade expansion through MRAs within the Asia-Pacific region represents a critical, new opportunity for export-led U.S. economic advance.

In order to facilitate U.S. and global trade expansion and economic advance, the following are recommended:

- **RECOMMENDATION 5:** The Office of the U.S. Trade Representative should continue ongoing mutual recognition agreement negotiations with the European Union. The USTR should also expand efforts to negotiate MRAs with other U.S. trading partners in markets and product sectors that represent significant U.S. export opportunities. Priority should be given to conclusion of MRAs for conformity assessment through the Asia Pacific Economic Cooperation forum.

- **RECOMMENDATION 6:** The USTR should use its authority under Section 301 of the Trade Act of 1974 to self-initiate retaliatory actions against foreign trade practices involving discriminatory or unreasonable standards and conformity assessment criteria. In particular, if U.S.-EU negotiations do not succeed within two years in securing fair access for U.S. exporters to European conformity assessment mechanisms, the USTR should initiate retaliatory actions under Section 301.

There are clearly problems of access for global producers in accessing national markets outside Europe where use of Section 301 may be warranted. The congressional request for this report, however, specifically referenced current EU-U.S. trading relationships and access for U.S. firms to European markets. Negotiations to remove barriers to U.S. manufacturers in Europe have been ongoing for a substantial period of time. It is reasonable, therefore, to promote closure of this serious, outstanding issue within a two-year time frame.

In addition to policies aimed at reducing trade barriers, innovative export promotion programs will also provide economic benefit. The United States has a unique opportunity to provide leadership, facilitate world trade, and promote U.S. exports through technical assistance to countries in emerging markets. U.S. government and industry should provide expertise and resources to assist developing countries in establishing open, fair standards and conformity assessment systems. As noted in Chapter 4, NIST is currently developing an expanded program on International Trade Standardization and Measurement Services that serves to meet some part of this overarching goal.[7] Programs such as these facilitate trade expansion. In addition, technical assistance has the potential to increase our exports by promoting adoption of both U.S. standards and international standards developed by U.S. industry as a means of meeting recipient countries' economic and regulatory objectives. The following is therefore recommended:

- **RECOMMENDATION 7:** NIST should develop and fund a program to provide standards assistance in key emerging markets. The program should have four functions:

    (a) provide technical assistance, including training of host-country standards officials, in building institutional mechanisms to comply with the Agreement on Technical Barriers to Trade under the Uruguay Round of the GATT;

    (b) convey technical advice from U.S. industry, standards developers, testing and certification organizations, and government agencies to standards authorities in host countries;

    (c) assist U.S. private-sector organizations in organizing special delegations to conduct technical assistance programs, such as seminars and workshops; and

    (d) report to the export promotion agencies of the Department of Commerce (such as the U.S. and Foreign Commercial Service) and the USTR regarding standards and conformity assessment issues affecting U.S. exports.

To accomplish the objectives of the program, NIST should station U.S. technical experts in key foreign markets, including the rapidly emerging economies of Asia and Latin America. These experts must have at least five years of directly relevant experience in standardization, testing, and certification. They should be recruited primarily from the private sector and accredited by NIST as U.S. government representatives.

MEETING FUTURE CHALLENGES

As noted in this report, data and analysis are lacking on the economic effects of domestic and international standards and conformity assessment systems. Success in efforts to improve U.S. systems will be greatly facilitated by increased federal data-gathering and analytical capacities. In addition, over the next several decades, there will be important new, international developments in such areas as environmental management process standards. These developments will confront industrial leaders and policymakers with a critical need for information, monitoring, and early analysis of global standards and conformity assessment issues.

Accessing information about international developments is especially difficult for small and medium-size U.S. firms. Future U.S. export expansion will be influenced by the ability of these firms to meet standards and certification requirements of their customers abroad, as well as those of domestic customers who incorporate their products into finished goods for export. The recent agreement between ANSI and NIST to establish a national electronic network to interconnect existing sources of standards information is a first step toward improvement of information dissemination.[8] Additional mechanisms must be established, however, to develop new sources of information and to reach small firms that lack access to electronic data networks.

In order to build a national capacity to monitor, assess, and disseminate information about international trends as they affect U.S. economic interests, the following are recommended:

- **RECOMMENDATION 8:** NIST should increase its resources for education and information dissemination to U.S. industry about standards and conformity assessment. NIST should develop programs focusing on product acceptance in domestic and foreign markets. These efforts should include both print and electronic information dissemination, as well as seminars, workshops, and other outreach efforts. Programs should be conducted both by NIST staff and by private organizations with NIST cooperation and funding.

- **RECOMMENDATION 9:** NIST should establish a permanent analytical office with economics expertise to analyze emerging U.S. and international conformity assessment issues. The office should evaluate and quantify the cost to U.S. industry and consumers of duplicative conformity assessment requirements of federal, state, and local agencies. To support the work of the USTR and other federal agencies, including those involved in export promotion, it should also collect, analyze, and report data on the effects of foreign conformity assessment systems and regulations on U.S. trade.

The need for information is acute in areas of emerging complexity in international conformity assessment systems, including third-party testing, certification, accreditation, and quality system registration. These systems have the potential for significant impact on U.S. manufacturers' costs and export opportunities. As noted in Chapter 3, for example, the rapid spread of requirements for certification of manufacturers' quality management systems to the International Organization for Standardization (ISO) 9000 series of standards was largely unanticipated in the United States. Registration of quality systems by third-party registrars imposes a significant cost burden on manufacturers, which is particularly difficult for small firms to bear. Furthermore it is important to recognize that ISO 9000 registration is related only indirectly to final product quality. Consistency and documentation of manufacturing processes, the focal points of ISO 9000, do not guarantee product quality if other elements, such as excellence in product design, are missing.

Mechanisms for auditing and certifying firms' environmental management systems, now under development in ISO, may follow a similar pattern. Without careful monitoring and active participation of U.S. government and industry experts in the early stages of development, environmental management standards have the potential to restrict trade and add unjustified complexity and cost to international conformity assessment systems, while providing only indirect enhancement of environmental protection. Anticipatory analysis of these and other emerging standards and conformity assessment issues is needed, therefore, to enhance U.S. trade and standards policymaking. The following is recommended:

• **RECOMMENDATION 10:** The USTR's post-Uruguay Round trade agenda, including work through the World Trade Organization, should include detailed analysis and monitoring of emerging environmental management system standards and their potential effects on U.S. exports. Technical assistance should be provided to USTR by NIST.

NOTES

1. See, for example, U.S. Congress, Office of Technology Assessment, *Global Standards: Building Blocks for the Future*. For a bibliography of economics literature on standards development, see Paul A. David and Shane Greenstein, *The Economics of Compatibility Standards: An Introduction to Recent Research*.

2. U.S. Environmental Protection Agency, Jeanne Hawkins, ed., *Final Report of the Committee on National Accreditation of Environmental Laboratories*.

3. As noted in Chapter Three, numerous organizations provide accreditation services in the private sector. These include, among many others: the American National Standards Institute (New York) which accredits product certifiers; the American Association for Laboratory Accreditation (Gaithersburg, Md.) which accredits laboratories for competence in specific fields of testing; and the Registrar Accreditation Board (Milwaukee, Wisc.) which accredits quality system assessors and registrars.

4. U.S. Congress. *Fastener Quality Act*, H.R. 3000.

     5.   Breitenberg, ed., *Directory of Federal Government Laboratory Accreditation/Designation Programs*, 52-53. Hank Woodcock, *Nationally Recognized Testing Laboratories*, and personal communication, Charles W. Hyer, Executive Vice President, The Marley Organization, March 4, 1994.

     6.   U.S. consensus standards-developing organizations were identified in Chapter Two. These include, for example, the American Society of Mechanical Engineers; the American Society for Testing and Materials; the Institute for Electrical and Electronics Engineers; the Computer and Business Equipment Manufacturers Association; and the National Fire Prevention Association, among others.

     7.   For an additional discussion by a NIST advisory panel of industry and academic experts of the need for an expanded NIST effort in foreign assistance related to standards, see Visiting Committee on Advanced Technology of the National Institute of Standards and Technology, *International Standards Issues: A Statement to the Secretary of Commerce*, p. 4-6.

     8.   For a program description, see American National Standards Institute, *1993 Annual Report*, 8-9. For additional discussion of the potential value of a standards information network, see Visiting Committee on Advanced Technology of the National Institute of Standards and Technology, *International Standards Issues: A Statement to the Secretary of Commerce*, p. 3-4.

Appendixes

APPENDIX

A

New Developments in International Standards and Global Trade: A Conference Summary

John S. Wilson, John M. Godfrey, Holly Grell-Lawe

On March 30, 1994, the International Standards, Conformity Assessment, and U.S. Trade Policy Project and the Academy Industry Program of the National Research Council convened a conference to explore new developments in international standards and global trade. Conference participants included leaders from industry, government, academia, and private-sector organizations. Participants discussed standards as technical barriers to trade and the economic benefits of international trade in the post-Uruguay Round trade environment; national standards and conformity assessment systems in overseas markets; the U.S. government role in standards and conformity assessment; and standardization in the information technology and telecommunication industries.

This appendix summarizes the conference presentations and discussions, highlighting key points and issues raised by the participants. It does not represent a consensus opinion of the panels or participants, nor does it contain policy recommendations.

THE POST-URUGUAY ROUND TRADE OUTLOOK: ECONOMIC BENEFITS OF INTERNATIONAL TRADE

From a government perspective, trade is a tool of domestic policy. Since

John S. Wilson is Project Director/Senior Staff Officer, National Academy of Sciences. John M. Godfrey is Research Associate, National Academy of Sciences. Holly Grell-Lawe is Senior Research Associate, Center for International Standards and Quality, Georgia Institute of Technology. Special thanks to Patrick Sevcik, Project Assistant, National Academy of Sciences, for assistance in the coordination of the conference and preparation of this report.

conclusion of the North American Free Trade Agreement (NAFTA) and the Uruguay Round of multilateral trade negotiations of the General Agreement on Tariffs and Trade (GATT), policies to reduce barriers to trade and expand the total volume of trade are among the principal tools available to enhance the long-term U.S. economic growth rate.

A senior U.S. government trade official noted the desire of Congress and the public for empirical evidence that trade liberalization agreements enhance long-term growth prospects and prosperity, and reported some empirical findings, which are outlined below:

The Office of the U.S. Trade Representative (USTR) examined the wage levels of U.S. export workers in 1990, building on work done by the U.S. Department of Commerce (DoC), which had calculated that 7.2 million jobs are supported by U.S. merchandise exports. Matching the DoC data with average wages per industry, USTR found that U.S. workers in export-related sectors earn 17 percent *more* than the U.S. national average wage.

The USTR also reviewed economic research on the wage levels of workers in import-competing sectors. One study found that U.S. workers in import-competing sectors earned 16 percent *less* than the U.S. national average wage. The general conclusion reached by many studies is that policies lowering barriers and expanding market-driven trade will gradually shift the growth of job opportunities from lower-paying jobs toward higher-paying jobs.

In connection with the Uruguay Round, the U.S. Bureau of Economic Analysis explored both dynamic (growth) effects and static efficiency effects from a one-third cut in global barriers to trade, a primary goal of the Uruguay Round negotiations. It was found that for the United States at the end of 10 years there would be a growth enhancement of about 3 percentage points of Gross Domestic Product. The U.S. trade official noted that few policy levers can add an annual *three-tenths* of a percentage point to the country's long-term growth rate.

Developing nations around the world are moving toward more market-oriented policies, both internally and in their trade policies. In doing so, they have the potential for real economic growth rates far exceeding anything achievable in the United States, the European Union (EU), or Japan. Most of these countries have very broad needs—from telecommunications systems to road-building equipment to hospital equipment. This presents a tremendous opportunity for U.S. exporters.

Standardization is one of many areas that must be addressed to enhance the U.S. ability to take advantage of these opportunities. Standards can facilitate U.S. exports. Standards, however, may also pose barriers to trade. The United States needs to decide how to deal with standards issues more fully with developed country trading partners. The United States also has an interest in ensuring that as developing countries are more integrated into the global trading system, they adopt open standards models in which government's ability to interfere and distort trade flows and reduce growth is limited and can be addressed through negotiation where it does exist.

THE POST-URUGUAY ROUND TRADE OUTLOOK: STANDARDS AND TECHNICAL BARRIERS TO TRADE

Significant changes in the global trading system have occurred over the past decade. The Uruguay Round of the GATT has been concluded, lowering tariff barriers to international trade in many industry sectors. Non-tariff barriers to trade, such as standards, however, have become a growing source of international friction. Standards have been identified by U.S. industry as a problem area in trade relations, which means standards have become increasingly important to the USTR.

Standards can serve as technical, non-tariff barriers to trade in several ways. In the setting of standards, it is possible that the standards adopted, their method of application, and the procedure to assess conformity will discriminate against foreign suppliers. Lack of access to information about what standards are and how they should be met also impedes global trade. Every country has the right to maintain its own standards, when necessary to protect health, safety, and the environment. Without harmonization of international standards wherever possible, however, differences in standards can result in barriers to trade. Trade is also restricted, and potential growth and economic welfare are reduced when there are (1) differences in standards, (2) needlessly burdensome conformity assessment processes, and (3) failure to reduce the costs of those processes by establishing methods and procedures to facilitate conformity assessment.

STANDARDS AND TECHNICAL BARRIERS TO TRADE: THE EUROPEAN AND GATT APPROACHES

Both the Single Internal Market program of the EU and the Uruguay Round have wrestled with the issue of standards as technical barriers to trade. As a result, the EU has adopted a comprehensive approach to standards harmonization among its member states. The Uruguay Round addresses this issue by establishing a code of rules for governments and standards-setting bodies.

Prior to 1985, standards were a major stumbling block to European integration into a single market. Standards in Europe were created by governments. Harmonization required governments to agree on details of a common approach, but this proved impossible. In 1985, however, a new approach emerged. The solution gave industry responsibility for the development of standards that would meet "essential safety and health requirements" mandated by EU legislation. This solution limited the role of the EU Commission to the definition, through legislation, of broad safety and health objectives to be met. Industry and private standards-setting bodies provide the details of how those objectives are to be achieved.

The Uruguay Round addresses standards at a global level by establishing a system of rules, the Agreement on Technical Barriers to Trade (TBT), by which government and private standards-setting bodies are to act in the future. In the view of one European panelist, this is a significant achievement of the Uruguay Round. Standards can serve as barriers to trade, and further problems can be anticipated over time, for example, as environmental concerns generate new standards and regulations.

Universality is another important achievement of the Uruguay Round TBT agreement. There are more than 115 signatories to the GATT, as amended, agreeing to abide by one comprehensive set of rules. A previous attempt in the Tokyo Round of the GATT to adopt a separate, optional technical barriers to trade code, the Standards Code, resulted in only 38 countries agreeing to exchange information on standards.

It is also important that the new TBT agreement, an integral part of the GATT, invokes a higher level of obligation. The signatories (sovereign governments) agree to take full responsibility in their respective countries for the application of the terms of the code to whatever level of government or private sector is involved.

In addition, both public and private standards-setting bodies are enjoined not to act in a discriminatory fashion or to use standards as hidden barriers to trade. It was noted that the EU was anxious to have the United States agree to this provision. In exchange, the EU agreed to submit itself to national treatment and be considered the responsible level of government. This will require all lower levels of government to adhere to the TBT. As one conference participant noted, the United States agreed to a certain degree of coverage of its private-sector standards bodies in return for more direct coverage through Brussels of European standards bodies, including the European Committee for Standardization (CEN) and the European Committee for Electrotechnical Standardization (CENELEC). In the view of a European panelist, this mitigates the danger of discriminatory application of the European standards system in the future.

With respect to government obligations concerning private standards bodies under the Uruguay Round agreement, there is a Code of Good Conduct under which private-sector bodies can submit to the Most Favored Nation Clause. If a private-sector body signs the code, it must treat others in a nondiscriminatory fashion. This means, for example, that if a standards organization admits some foreign participation in its procedures, it may not block other foreign participants.

Another achievement of the Uruguay Round has been to put in place not only agreements on technical standards, but also a multilateral trade body, the World Trade Organization. The dispute settlement system of this multilateral trade organization applies directly to technical standards as barriers to trade. Since the GATT is an intergovernmental organization, private organizations or firms can access the dispute settlement system only through their government. On the other hand, the development of private diplomacy among standards-setting bodies at a global level is possible and encouraged in the text of new international trade agreements.

U.S. industry participants expressed concern about the dispute settlement system under the GATT. Improving the dispute settlement mechanism and integrating it into the Uruguay Round represented a very high priority for the United States. It was pointed out, however, that how often the dispute settlement mechanism will be invoked with respect to specific unfair trade practices is an open question.

The USTR was encouraged to draw on the advice and resources of private companies in bringing cases to the GATT. With respect to dispute settlement, the TBT agreement will have no practical effect unless companies bring cases to the GATT. The USTR requires this kind of assistance. It was also pointed out, however, that the GATT dispute settlement procedure is the end of a long process. Most disputes should not end up in the formal dispute settlement procedure, but should be settled privately.

The topic of access to the European standards system by non-European private-sector standards bodies elicited considerable discussion. The EU and Uruguay Round texts enjoin the private standards-setting bodies to participate actively in horizontal communication between groups, as well as in the International Organization for Standardization (ISO) and its subgroups. A European panelist noted that the easiest way to gain access to the European standards-setting system is via the ISO and strongly urged that U.S. firms (1) work to ensure full representation in the work of the ISO and (2) work to establish international standards. Where international standards exist, there is a general obligation, in the EU and GATT texts, for standards bodies to incorporate them in their own standards to the greatest possible extent.

The importance of the private sector in the standards process was highlighted during conference discussions. In the view of a European participant, if there is a full commitment in the United States and in Europe to develop international standards, the new processes outlined in the Uruguay Round will function properly. If there is no private-sector commitment, they will not work. The private sector needs to be an active participant where its interests lie. The hope was expressed that the idea that domestic markets can be protected with standards will be left behind as we enter the twenty-first century.

The emphasis on participation in international standards for access to European standards setting is not new. For example, as one U.S. industry participant commented, over a period of four years, the American National Standards Institute (ANSI) has had a highly productive dialogue with CEN, CENELEC, and the European Telecommunications Standards Institute. ANSI receives these organizations' standards for comment and has an opportunity to comment on their processes. The challenge has been to motivate U.S. industry to use these opportunities for access more effectively.

Only U.S. multinational firms with operations in Europe have *direct* access to CEN and CENELEC committees. The answer to the problem of access, according to a European participant, is for private standards-setting bodies to develop international standards. It was noted that the United States does participate in many areas of international standards activities. For example, the United States holds 35 technical committee secretariats in the International Electrotechnical Commission (IEC) and is the largest single holder of secretariats in ISO.

NATIONAL STANDARDS AND CONFORMITY ASSESSMENT SYSTEMS IN OVERSEAS MARKETS: A U.S. PERSPECTIVE

In the early part of the twentieth century, the world marketplace comprised strong, individual, independent trading markets. It now comprises a number of economically strong nations and trading regions. The integration of international markets has impacted product certification and standardization activities in a number of ways. Important changes are taking place in standards and conformity assessment systems in Europe, Asia, and emerging markets, affecting both regulated and unregulated product sectors.

Worldwide, standards harmonization activities have increased. For example, the countries of the EU and the European Free Trade Association (EFTA) are involved in harmonization efforts, while Japan is examining international standards harmonization, particularly as it relates to quality. Under NAFTA, Mexico is harmonizing its standards with the United States and Canada. Product certifiers involved in standards have increased their participation in international activities and standards harmonization. One U.S. organization participates in more than 150 international standards committees and 60 harmonization initiatives around the world.

A conference panelist from a U.S. testing laboratory reported that in response to changes in international trade, product testers and certifiers in the United States and other developed markets are internationalizing their activities and introducing new services to meet changing international service demands of their customers. The evolution of global manufacturing gives certifiers incentives to develop a global presence.

The need for multiple market access has led to increased cooperation between certifiers through the establishment of memoranda of understanding, bilateral agreements, and other arrangements. These have built confidence in a capacity to certify, test, quality register, and provide other elements of conformity assessment services worldwide. U.S. certifiers have built agreements with certifiers in countries that are primary markets for U.S. products.

The experience of one U.S. product certifier suggests that the keys to cooperation in assessing and certifying products locally or supplying data locally in foreign markets, are confidence building, cross-training, audits, and the right to reject.

Certifiers can also provide global assistance to customers that want to export to new markets by identifying the requirements, codes, standards, and laws that a product must meet, including the processes by which a product must be evaluated. In the view of one participant, the certifiers of the future will need not only experience and flexibility, but also foreign market intelligence, transparency in foreign standards and certification processes, and the backing of national treatment before they can help producers reach U.S. export markets.

NATIONAL STANDARDS AND CONFORMITY ASSESSMENT SYSTEMS IN OVERSEAS MARKETS: A EUROPEAN PERSPECTIVE

A European perspective on conformity assessment was provided by a panelist from the European testing and certification community. Conference participants were advised that a new industry is developing around conformity assessment. There is the danger that it could develop in such a way that customers will reject it as an obstruction and try to find a way around it, at either the national, the European, or the international level.

Via the European Organization for Testing and Certification (EOTC), Europe is attempting to harmonize accreditation in nonregulated industry sectors through the calibration, certification, and laboratory accreditation communities in 18 countries. A private-sector, not-for-profit association, the EOTC seeks to establish confidence, through mutual recognition agreements, among parties concerned with conformity assessment issues.

To accomplish this, the EOTC uses *sectoral committees* and *agreement groups* that represent the interests of manufacturers, users, and third parties in a specific economic sector, and of the laboratory, certification, and accreditation communities. Comprising national delegations from at least five EU or EFTA countries, sectoral committees discuss mutual recognition arrangements where they perceive a market need for harmonized conformity assessment. Agreement groups are formed among organizations from at least three EU or EFTA countries that have agreed to operate in conformance with similar European or international standards. This requires that the national conformity assessment accreditation system be available to both domestic and foreign laboratories. One panelist noted that such systems are not presently in place in some European countries.

EOTC is working to promote conformity assessment arrangements whereby a product is tested once, certified, and accepted everywhere within the market, whether in Europe or in another country through a mutual recognition arrangement. It was noted that harmonized or compatible conformity assessment arrangements within the EU or other countries would remove the requirement that U.S. products be sent to European laboratories or certification bodies for tests.

In the discussion of conformity assessment in overseas markets, several U.S. industry participants stressed the need to maintain reliance wherever possible on self-certification, by using the supplier's declaration of conformity rather than services provided by an independent third party.

THE ADMINISTRATION'S DEFENSE ACQUISITION REFORM POLICIES

An overview of the recommendations of the U.S. Department of Defense (DoD) Process Action Team (PAT) for Specifications and Standards was provided by a conference participant from DoD.

The PAT was chartered to, among other tasks, develop a comprehensive strategy for permitting reliance in defense procurement on commercial products, specifications, and standards, as well as practices and processes. The team works to ensure that (1) system and data requirements do not unnecessarily preclude the use of commercial products and practices in form, fit, and function; (2) unnecessary specifications and standards are eliminated; (3) nongovernmental standards and commercial item descriptions are used to the maximum extent practical; (4) industry is encouraged to propose alternative technical solutions; and (5) standards and specifications are applied correctly to contracts.

Several of the team's primary recommendations include the following:

• *Performance specifications*: For all Acquisition Category programs for new systems, major modifications, technology generation changes, nondevelopmental items, and commercial items, DoD's need shall be outlined in terms of performance specifications. The contractor, rather than DoD, will be responsible for designing and developing solutions to meet these performance requirements.

• *Manufacturing and management standards*: Management and manufacturing standards shall be canceled or converted to performance or nongovernment standards. Contractors will be given the option, in complying with military specifications (milspecs), of proposing relevant nongovernmental standards or industry practices as substitutes that meet the intent of the specific milspecs or military standards.

• *Innovative contracting processes*: All new high-value solicitations and ongoing contracts will include a statement encouraging contractors to submit alternative solutions to milspecs and standards. Incentives will be provided to allow current contractors to switch to a new specification or standard.

• *Partnerships with industry associations*: DoD will encourage partnerships with industry to help replace milspecs and standards with nongovernmental standards, where practical.

• *Reduction of oversight*: DoD's recent directive authorizing the use of the

ISO 9000 series of quality management system standards allows, but does not mandate, the use of the series. DoD will no longer dictate how contractors establish quality systems. Contractors will be responsible for establishing systems, processes, and standards.

 • *Establish goals to reduce the cost of contractor development and production test and inspection*: The team recommended that DoD set a specific goal of reducing the cost of contractor development, production, testing, and inspection, as well as the use of innovative techniques utilized in industry. These include simulation, environmental testing, dual-use test facilities, process controls, metrics, and continuous process improvement.

It was also noted that in the acquisition of commercial products, DoD will rely on third-party certification of compliance with standards in place of DoD certification where possible. Where such programs do not exist, DoD will cooperate with industry to establish them. DoD is working to make maximum use of third-party organizations.

It was acknowledged by a DoD participant that it will be difficult to change the culture in DoD and reengineer the acquisition process. The key to this effort is leadership in management. Metrics of success developed by teams will help determine both the progress and the impact of changes.

THE U.S. GOVERNMENT ROLE IN STANDARDS AND CONFORMITY ASSESSMENT

The U.S. government plays an active role in many aspects of standards and conformity assessment. Participants discussed existing and potential missions for the U.S. government in these areas. Also considered were the desirable mix of government and private-sector activity in standards and conformity assessment and ways to arrive at those desired relationships.

A Government Agency Perspective

In the changing international trade environment, according to a U.S. government panelist, many of our previous assumptions that others will generally adopt U.S. standards and buy U.S. products are no longer valid. In addition, international standards are often developed without active U.S. input or representation. Exporters also face aggressive trade practices by other countries, including the use of standards as barriers to trade. On the positive side, international cooperation is increasing as evidenced by both GATT- and NAFTA-related activities.

From the perspective of one U.S. government agency, the private voluntary standards system represents the best approach for the U.S. economy. There is a need, however, for much better cooperation and communication among standards organizations, industry, and government to make this system work effectively.

The U.S. government's general role in standards is, first, as a user of standards. Agencies use standards in the purchase of products (e.g., DoD) and through their incorporation into federal regulations (e.g., the Occupational Safety and Health Administration). The government also provides the technical foundation for many standardization activities and advocates for U.S. interests around the world.

The National Institute for Standards and Technology (NIST) plays many roles in standardization. NIST provides the technical basis for standards through fundamental physical standards measurement, test methods, reference data, and production of standard reference material. NIST participates in voluntary standards committees, presently holding 1,143 memberships on 816 standards committees of 79 organizations. NIST chairs 118 of these standards committees. NIST also runs the National Voluntary Laboratory Accreditation Program. In addition, about 170,000 standards-related inquiries are handled annually by NIST's National Center for Standards and Certification Information. The center serves as the U.S. GATT Inquiry Point. NIST has begun to explore mutual recognition of accreditation bodies through its National Voluntary Conformity Assessment Systems Evaluation program. Finally, NIST is working to improve communication and cooperation among government agencies, standards organizations, and industry.

The U.S. standards system faces several challenges. The current process may not always be adequate to deal with the changing international environment. For example, ANSI is often the U.S. representative in the international standards arena. In one panelist's view, ANSI's lack of formal government backing can place it at a disadvantage relative to other parties. The United States often acts in a reactive mode in some areas of standards, rather than setting the agenda for the rest of the world. There is some movement by the private sector, however, to recognize the importance of participation in the international standards arena. This is evidenced by the growing number of U.S.-held ISO committee chairs.

A government participant proposed developing a systems approach to the standards process to achieve national goals more effectively. This would require (1) focusing on a clearly defined national goal; (2) delineating responsibilities and relations among standards organizations, industry, and government; and (3) improving communication among all parties in the voluntary standards process.

A Private-Sector Perspective

A private-sector panelist's perspective on the U.S. government's role in standards and conformity assessment emphasized the need for a more definitive relationship between the government and the private voluntary standards system. The panelist stressed that government must recognize and use the private voluntary standards system. In addition, U.S. industry should support ANSI as a credible mechanism for promoting integration of the voluntary standards system.

The U.S. government must also organize to speak coherently as a sovereign nation on the issues of standards and conformity assessment, both internally and externally. This is particularly important in view of the new responsibilities and accountability of the sovereign government signatories to the Uruguay Round. Finally, the cost of standardization should be borne by those who gain value from the activity.

Many conference participants agreed that the United States must support an infrastructure that is competent, equitable, and credible. Some expressed the view that ANSI and the voluntary standards system should not be governed and funded by the present mixture of government, associations, standards developers, and a small percentage of industrial firms in the United States. This represents an uneven and inequitable distribution of governance and of financial support. A private-sector panelist maintained that revenue must be derived from value received and that ANSI should concentrate on activities that are of value, identifying who values them, and assessing revenue sources from that value structure.

The future will be substantially different from the past. Both private and public leaders need to seek improvements to the U.S. standards system in a way that involves the government. One participant pointed out that undesired levels of government control, which may accompany government involvement in private activities, can be mitigated through negotiation.

Industry–Government Cooperation in Regulation and International Standards

The medical device industry and the Food and Drug Administration (FDA) have developed and implemented a program to provide industry access to effective mechanisms to influence international standards, regulations, and trade. This program has been built on cooperation between industry and government. In 1989, the U.S. medical device industry instituted a program to influence international and European Community standards to promote harmonization of medical device standards and regulations. The industry administers most national standards and almost all horizontal medical device international standards.

The program's success could not have been achieved without the support of the FDA, according to an industry panelist. The program was enhanced by the fact that the U.S. industry's standards were used by many foreign governments. With FDA's support, the industry adopted as national standards all of the international standards that it administers in the United States. FDA support was critical to the medical device association's involvement with an ISO technical committee that will write, interpret, and coordinate medical device quality system standards for the international community. It is also working to harmonize medical device standards and regulatory requirements for all countries worldwide.

FDA provided leadership, resources, and credibility to this exercise. FDA also provided scientific and political leadership to industry's efforts to influence

ISO, IEC, CEN, and CENELEC standards. Through the FDA, the industry gained access to CEN and CENELEC committees. The U.S. medical device industry was able to join standards-writing committees by gaining the respect and confidence of the Europeans and making it clear to them that U.S. industry has as much interest in writing appropriate standards as they have.

The FDA has also assisted industry leaders in promoting the message that international standards activities are important, not only to industry but to the FDA. U.S. industry in general needs to make more of a commitment to international standards, and more companies should carry the burden, emphasized the industry panelist.

The FDA has benefited from this activity in several respects. First, FDA relies on standards for regulatory activities, almost all of which are based on some form of voluntary standards. These include manufacturing practices, product approval processes, and clinical investigations. FDA experts are also provided access to experts from all over the world. As a technical regulatory body, FDA needs to maintain its knowledge base. The most important reason for FDA's interest in international standards may be that harmonization of international standards and regulations can significantly reduce FDA's regulatory cost.

In the view of one industry participant, it is critical that FDA continue as a major participant in the international standards system. Its leadership, knowledge, resources, and management can contribute to industry's effort to ensure that standards expand global market access rather than limit it. For example, FDA experts support the formulation and acceptance of U.S. positions on international standards. Continued funding for participation in these international activities was encouraged by the industry panelist.

Finally, the medical device industry supports the notion that the government should be a full and equal participant in, but not the director of, private-sector international standards participation. It is important for the government and the private sector to speak with a unified voice. FDA has shown a willingness to share responsibility with the private sector in international standards, commented an industry participant.

Standardization and harmonization processes in sectors other than medical devices can likely be furthered by cooperative relationships between industry and government. For some 25 years, industry—working in cooperation with the FDA rather than being directed by it—has produced a tremendous number of standards critical to FDA regulation. It also has made a major contribution to marketplace safety.

Comments and Concerns

A number of other comments and concerns about the U.S. government role in standards and conformity assessment were voiced by conference participants. Several of these are highlighted below:

There are several existing mechanisms that demonstrate significant interface among the U.S. government, industry, and standardization matters. One is the Government Member Council of ANSI, and another is the NIST-chaired Interagency Committee for Standards Policy.

There is a need for an exchange of information concerning standards developments among federal agencies. Another challenge is overall coordination of the government's role in standards, by taking into account the differing roles and missions of the various agencies involved in standards-related matters.

Some participants maintained that ANSI would benefit from a federal charter, while others did not see the need for the additional processes or potential bureaucracy that might result from such formalization.

The government can assume a more proactive role in standards by working to influence standards setting in developing countries or in emerging markets through the provision of technical support or training. Europe and Japan are active in this area. It was noted that NIST has such an effort ongoing with Saudi Arabia.

SECTORAL CASE STUDIES: TELECOMMUNICATIONS AND INFORMATION TECHNOLOGY

Standards issues are an increasingly important and contentious issue in the telecommunication and information technology industries. This has been driven by several forces. First, the number of players in the standards process has increased markedly as government policies have opened many markets to new entrants. In addition, the rate at which these industries need to develop new standards is accelerating as rapid technological progress leads to the introduction of both new services and new ways of providing old ones. The blurring of traditional industry lines in these sectors is also driving demand for changes in standards as firms capitalize on the convergence of industries and technologies to create advanced services for customers. Finally, as many firms have moved outside traditional domestic markets to compete internationally, foreign standards have become important to their success and the role of international standards has become more critical.

Telecommunications

The telecommunication industry is dependent on universality, which depends in turn on standards. In the wake of the dynamics mentioned above, it is critical that standards become adaptable at a much faster pace, according to a telecommunication industry executive. The key is to develop standards that from the outset adapt to the movement of the industry. To accomplish this objective, standards need to be open, scalable, and extensible: open in terms of being available to all potential providers on a equitable basis; scalable in that they can adapt readily to improve performance in a given technology; and extensible to new technologies as they become available.

In evaluating the telecommunication standardization process, the question of whether formal standards are necessary arises. In some cases, *not* developing formal standards could lead to chaotic marketplaces or to inefficient or costly services. In other cases, standards may be better left to the marketplace. In yet other cases, adoption of standards could be counterproductive.

The government should not actually set standards, in the view of the telecommunication panelist. The government does not generally have the prerequisite technical expertise or marketplace savvy. Government can, however, encourage and participate in the standards process in order to meet public interest responsibilities. It was noted that the private sector is motivated to get involved in the standards process when adopting a common standard would produce a far larger market with less risk of failure than would a market consisting of multiple incompatible technologies.

Historically, most telecommunication standards bodies were *not* formed with the notion of creating standards that would invite open competition and rapid change. Instead, they were formed under the aegis of national monopoly service providers. Although telecommunication standards processes work, they have been generally slow. The implementation of accelerated procedures has improved the speed of the processes, but not enough, in the opinion of a telecommunication industry participant.

Rapidly changing telecommunication technologies and markets should force industry consideration of some fundamental goals: (1) acceleration of the standards process; (2) attention not only to the most current technology, but to technologies that can be anticipated and technologies that are available to competing countries; and (3) making standards as flexible, scalable, and extensible as possible, by anticipating improvements in present and future technologies.

Information Technology

Standards drive development in the information technology (IT) industry. Customers demand interoperability among networks, computers, applications, and people, as well as the freedom to choose between vendors. The IT industry has a broad range of formal and informal standards development processes. The U.S. industry has been successful in having its standards adopted internationally through private-sector processes.

Informal systems for IT standards development include a variety of consortia. The industry has formed multiple consortia to perform specific technical work, promote standards and ensure their implementation, test products, and build common software. Consortia are usually formed, however, to facilitate the timely completion of standards and ensure that they pertain to real products.

Consortia have done effective and efficient work in regard to the latter. However, there is no credible evidence that the consortium-based process is inherently faster than the formal standards process, according to an IT industry panelist. In particular, consortia sometimes fail to meet their objectives because they believe they can short-circuit the painful political process of developing consensus on a broad base. Relying on the willingness of the participants to passively follow the major players has not proved an ideal solution.

Consortia have nevertheless remained popular because (1) they are involved in activities other than standards development and (2) they focus on near-term success by concentrating on existing technology and product implementations. Consortia also appear to use effective methodologies. They tend to work more intensively than formal standards groups. They make better use of technology and therefore have the potential to be more efficient. Yet, unlike the formal standards process, consortia may exclude the small participant because of the higher level of resources required to participate.

Consortia will continue to be a factor in the IT industry. They add value and are easy to form. One problem, however, is that the number of groups has increased so dramatically that there is considerable confusion about which standards are the best technical solution. It has become difficult to judge the credibility of any particular standard or standards group. There is also a general inability to deal with mutually exclusive standards. Vendors react to this situation by participating in all the groups and implementing all standards. Users express their frustration and demand the formation of yet another new consortia to do it right. This results in inefficiency and added costs.

There is a need to take advantage of the strengths and weaknesses of all these groups. Actions suggested by an industry panelist include, first, recognizing a preference for the formal processes at the national and international levels. At the same time, the formal standards process needs to separate technical development from formal approval. In this way, the IT industry can take advantage of good work being accomplished, while maintaining a system for establishing legitimacy and minimizing confusion. This will require changes in the standards process. One possibility would be to give international consortia nonvoting membership in international organizations.

With respect to the government role in IT standards processes, several suggestions were made. These included (1) continued active participation in the voluntary standards process on the same basis as other interested parties; (2) a greater funding role through, for example, coverage of specific infrastructure items (e.g., ISO and IEC dues, hosting of international standards meetings in the United States); (3) R&D tax credits for standardization activities; (4) negotiation of mutual recognition agreements, where required; (5) offering such services as accreditation of laboratories or recognition of private-sector accreditation programs; and (6) formal recognition of the existing private-sector standards system through, for example, the granting of a federal charter to ANSI.

NEW DEVELOPMENTS IN INTERNATIONAL STANDARDS AND GLOBAL TRADE

International Standards, Conformity Assessment,
and U.S. Trade Policy Project Committee
and
The Academy Industry Program

Wednesday, March 30, 1994

National Academy of Sciences Building
Washington, D.C.
The Lecture Room

- Agenda -

8:30-9:30 a.m.	Continental Breakfast and Registration
9:30-9:40 a.m.	*Welcome*

STEPHEN A. MERRILL, Executive Director, Science, Technology, and Economic Policy Board Director, Academy Industry Program

Introduction

JOHN SULLIVAN WILSON, Study Director, International Standards, Conformity Assessment, and U.S. Trade Policy Project

9:40-10:20 a.m.   *The Post-Uruguay Round Trade Outlook: Standards and Technical Barriers to Trade.* The panel will address the status of standards and conformity assessment systems as barriers to trade in the post-Uruguay Round trading system.

Moderator: Gary Hufbauer, Senior Fellow, Institute for International Economics

David Walters, Acting Assistant U.S. Trade Representative for Economic Affairs and Chief Economist, Office of the U.S. Trade Representative

Victoria Curzon-Price, University of Geneva

10:20-10:50 a.m.	Open discussion: Uruguay Round and changes to technical trade barrier provisions—effects on U.S. firms' technology development and export strategies; industry perspective on emerging standards-related trade issues, such as environmental standards, quality standards, and intellectual property protection
10:50-11:00 a.m.	Break
11:00-11:50 a.m.	*National Standards and Conformity Assessment Systems in Overseas Markets.* The panel will address trends in national standards and conformity assessment systems in prominent and emerging U.S. export markets.
Moderator:	Lawrence Wills, Director of Standards, IBM Corporation
	David Stanger, Secretary General, European Organization for Testing and Certification
	Laszlo Belady, Chairman, Mitsubishi Electric Laboratories
	S. Joseph Bhatia, Vice President, External Affairs, Underwriters Laboratories
11:50-12:30 p.m.	Open discussion: impact of standardization and conformity assessment trends in U.S. trading partner countries on U.S. exports; firms' experiences and participation in standards development in these countries
12:30-1:30 p.m.	Lunch
1:30 p.m.	*The Administration's Defense Acquisition Reform Policies*
	Coleen Preston Deputy Under Secretary for Acquisition Reform, U.S. Department of Defense
2:00-2:50 p.m.	*The U.S. Government Role in Standards and Conformity Assessment.* The panel will discuss current roles and potential new initiatives for the federal government in standards development, conformity assessment, and quality assurance.

Moderator:	Richard Schulte, Senior Vice President, Laboratories, American Gas Association
	Belinda Collins, Program Analyst, National Institute of Standards and Technology
	Robert Hermann, Senior Vice President, Science and Technology, United Technologies Corporation
	Michael Miller, President, Association for the Advancement of Medical Instrumentation
2:50 p.m.	Open Discussion
3:20 p.m.	Break
3:30-4:20 p.m.	*Sectoral Case Study: Information Technology (IT) and Telecommunication.* The panel will discuss the link among standards, technology, and development of new products and services in the context of rapid technological change and the ongoing convergence of these industry sectors.
Moderator:	Stanley Besen, Vice President, Charles River Associates
	Richard Liebhaber, Chief Strategy and Technology Officer, MCI Communications Corporation
	Stephen Oksala, Director, Corporate Standards, Unisys Corporation
4:20-4:50 p.m.	Open discussion: emerging trends in standards and technology development in the IT and telecommunication industries; role of international standardization in strengthening U.S. export performance
4:50 p.m.	*Concluding Remarks* GARY HUFBAUER
5:00 p.m.	Reception

PARTICIPANTS LIST

Melvyn R. Altman
Associate Director
Office of Standards and Regulations
Center for Devices and Radiological
  Health
U.S. Food and Drug Administration

Guy A. Arlotto
Deputy Director
Office of Nuclear Material Safety and
  Safeguards
U.S. Nuclear Regulatory Commission

Diana Artemis
Associate Director
Chemical Manufacturers Association

Daryl Back
Counsellor, Industry, Science and
  Technology
Embassy of Australia

Earl S. Barbely
Director, Telecommunication and
  Information Standards
CIP Bureau
U.S. Department of State

James Baskin
Director, Standards
NYNEX Corporation

Cora Beebe
Office of Management and Budget
Executive Office of the President

Laszlo Belady
Chairman
Mitsubishi Electric Research
  Laboratories, Inc.

Diego Betancourt
Manager
Office of Strategic Standardization
Polaroid Corporation

S. Joseph Bhatia
Vice President, External Affairs
Underwriters Laboratories, Inc.

Eric Biel
Trade Counsel
Senate Finance Committee

Ezra L. Bixby
President
Lovell Associates

Judy P. Boehlert
Vice President, Quality Control
Roche Pharmaceuticals
Hoffmann-La Roche Inc.

Barbara Boykin
Standards Coordinator
American Petroleum Institute

Richard Bradshaw
Vice President
North Atlantic Research, Inc.

Maureen Breitenberg
Economist
National Institutes of Standards and
  Technology
U.S. Department of Commerce

Dennis Brining
Director, International Program
  Development
Lockheed Corporation

George Brubaker
Deputy Director
Office of Standards and Regulations
U.S. Food and Drug Administration

Milton M. Bush
Director of Public Affairs
American Council of Independent
Laboratories

Amy Cheng
Industrial Engineer
Facilities Management Division
U.S. General Services Administration

Colin B. Church
Voluntary Standards and
International Activities Coordinator
U.S. Consumer Product Safety
Commission

Belinda Collins
Program Analyst
National Institute of Standards and
Technology

James D. Converse
Director of Corporate Standards
Eastman Kodak Company

Hugh Conway
Director
Office of Regulatory Analysis
Occupational Safety and Health
Administration
U.S. Department of Labor

Stephen Cooney
Senior Policy Director, International
Investment and Finance
National Association of
Manufacturers

Lori L. Cooper
Director, Internal Market Staff
Office of European Community
Affairs
U.S. Department of Commerce

Francis L. Criqui
Manager, Technical Document
Management
General Motors Corporation

Helen Davis
Washington Representative
American Society for Testing and
Materials

Lester Davis
Chief Economist
Department of Commerce

Myles Denny-Brown
International Economist
U.S. Department of Commerce

Helen A. Domenici
Associate, Federal Relations
Corporate Affairs Division
Pfizer Inc.

John Donaldson
Chief, Standards Code and
Information Program
National Institute of Standards and
Technology

Helen Donoghue
First Secretary
Washington Delegation of the
European Commission

Anita Drummond
American Council of Independent
Laboratories

Lester F. Eastwood
Director, Standards Strategy
Motorola, Inc.

Williams Edmunds
Manager, Codes and Standards
Owens Corning Fiberglas

George L. Edwards
President
Alliance for Telecommunications
    Industry Solutions

Thomas Ehrgood
International Trade Counsel
Digital Equipment Corporation

Fara Faramarzpour
Director
Office of Strategic Standardization
Polaroid Corporation

Wendell Fletcher
Senior Associate
Office of Technology Assessment

Thomas B. Fowler
Principal Engineer
The MITRE Corporation

Kim Frankena
Chief, Major Trading Nations Branch
U.S. International Trade Commission

James French
Director of Standards
American Institute of Aeronautics and
    Astronautics

Louis Geoffrion
Manager, Corporate Quality
    Assurance
Raytheon Company

Richard B. Gibson
Technical Standards Director
AT&T Bell Laboratories

Neal Goldenberg
U.S. Department of Energy

Melvin R. Green
Associate Executive Director, Codes
    and Standards
American Society of Mechanical
    Engineers

Alf S. Gunnersen
Associate Technical Director
The MITRE Corporation

Joseph K. Haeglin
General Manager, Central
    Engineering and Purchasing
Texaco Inc.

William F. Hanrahan
Senior Director
Computer and Business Equipment
    Manufacturers Association

William Hendrickson
Senior Editor
Issues in Science and Technology

Robert J. Hermann
Senior Vice President, Science and
    Technology
United Technologies Corporation

Peter L.M. Heydemann
Director, Technology Services
National Institute of Standards and
    Technology

Derek Holden
Vice President, Industry and
    Government Relations Market
    Development
Owens Corning Fiberglas Corporation

Virginia A. Huth
Office of Management and Budget
Executive Office of the President

Charles W. Hyer
Editor/Publisher
The Marley Organization, Inc.

Brian Kahin
Director, Information and
    Infrastructure Project
Science, Technology and Public
    Policy Program
John F. Kennedy School of
    Government
Harvard University

Sheldon Kimmel
U.S. Department of Justice

Louisa Kock
Program Examiner
Office of Management and Budget

Roy E. Lancraft
Division Manager
United Parcel Service

Holly Lawe
Senior Research Associate
Georgia Institute of Technology
Center for International Standards and
    Quality

Mary Anne Lawler
Director of Standards Relations
IBM Corporation

William Lehr
Assistant Professor
Graduate School of Business
Columbia University

Walter G. Leight
Deputy Director, Office of Standards
    Services
National Institute of Standards and
    Technology

Richard Liebhaber
Chief Strategy and Technology
    Officer
MCI Communications Corporation

Henry Line
Director, Global Products Standards
AMP Incorporated

John W. Locke
President
American Association for Laboratory
    Accreditation

Charles Ludolph
Director
Office of European Communities
    Affairs
U.S. Department of Commerce

Mark Mahaney
Council on Competitiveness

William A. Maxwell
Director, Government Relations
Convex Computer Corporation

Leroy M. May
Senior Staff Member and Standards
    Consultant
AT&T

Sergio Mazza
President
American National Standards Institute

Nina I. McClelland
Chairman, President, and Chief
    Executive Officer
NSF International

Mary McKiel
National Coordinator, EPA Standards
    Networks
Office of Pollution Prevention and
    Toxics
U.S. Environmental Protection
    Agency

Barbara McLennan
Staff Vice President, Government and
    Legal Affairs
Electronic Industries Association

Kenneth McLennan
President
Manufacturers' Alliance for
    Productivity and Innovation

Richard Meier
Deputy Assistant U.S. Trade
    Representative for GATT Affairs
Office of the U.S. Trade
    Representative

Michael J. Miller
President
Association for the Advancement of
    Medical Instrumentation

Thomas P. Monkus
Compliance Officer, Director QA/RA
Medical High Technology
    International, Inc.

Keith A. Mowry
Manager, Governmental Affairs
Underwriters Laboratories, Inc.

Robert W. Noth
Manager, Engineering Standards
Deere & Company

Stephen Oksala
Director, Corporate Standards
Unisys Corporation

Anthony R. O'Neill
Chairman of the Board
American National Standards Institute

J. Paul Oxer
Executive Assistant to the President
Ecology and Environment, Inc.

John P. Palafoutas
Director of Federal Relations
AMP Incorporated

Bernard J. Phillips
Head of the Competition and
    Consumer Policy Division
Organization for Economic
    Cooperation and Development

Herbert Phillips
Vice President, Engineering
Air-Conditioning and Refrigeration
    Institute

Kyle Pitsor
Manager, Energy and Trade
National Electrical Manufacturers
    Association

George Porter
Quality Systems Manager
ISO 9000
Roche Pharmaceuticals
Hoffmann-La Roche Inc.

Joseph Potts
Director
Technical Support Group
GTE Laboratories

Ernest H. Preeg
William M. School Chair in
    International Business
Center for Strategic and International
    Studies

Colleen Preston
Deputy Under Secretary for
    Acquisition Reform
U.S. Department of Defense

Victoria Curzon-Price
Professor of Economics
University of Geneva

Anne Rafferty
Economist
U.S. Department of State

Kim Ritchie
AVX Corporation

Sadi Ubaldo Rife
Counselor, Science & Technology
    Attache
Embassy of Argentina

Lloyd Rodenbaugh
Promega

Richard Rounsevelle
Chief, Recreational Boating Product
    Assurance
U.S. Coast Guard

Deborah Rudolph
Manager, Technology Policy
Institute of Electrical and Electronics
    Engineers - USA

William A. Ruh
Associate Technical Director
The MITRE Corporation

Francine Salamone
Associate Director, Medical
    Operations
International Pharmaceuticals Group
Pfizer Inc.

Andrew Salem
Staff Director
Institute of Electrical and Electronics
    Engineers

Gregory E. Saunders
Acting Director for Manufacturing
    Modernization
Office of the Secretary of Defense
U.S. Department of Defense

Mary H. Saunders
Assistant to the Director
Office of Standards Services
National Institute of Standards and
    Technology

Jane W. Schweiker
Director of Government and
    Organization Relations
American National Standards Institute

C. Ronald Simpkins
Manager, Corporate Engineering
Standards
E.I. DuPont DeNemours & Company

Anna Slomovic
Senior Policy Analyst
Rand Critical Technologies Institute

Oliver Smoot
Executive Vice President
Computer and Business Equipment
Manufacturers Association

Carlos Souto
Industrial Scientific Counselor
Embassy of Portugal

Michael B. Spring
Assistant Professor
Department of Information Science
University of Pittsburgh

David Stanger
Secretary General
European Organization for Testing
and Certification

Steve Stewart
Program Administrator, Public
Affairs Telecommunications
IBM Corporation

Gene Strull
Consultant
Westinghouse

Marty Sullivan
Director, Standards
Bellcore

Audrey Talley
Office of Food Safety and Technical
Services, Foreign Agriculture
Service, ITP
U.S. Department of Agriculture

John C. Tao
Director, EH&S
Audits Department
Air Products & Chemicals, Inc.

Keith B. Termaat
Executive Engineer, Engineering
Materials and Standards
Ford Motor Company

James Thomas
President
American Society for Testing and
Materials

Marie Thursby
Professor and Director
Center for International Business
Education and Research
Purdue University

Maria Tilves-Aguilera
Manager, Government Relations
Northern Telecom, Inc.

Robert Toth
President
R.B. Toth Associates

Suzanne Troje
Director, Technical Trade Barriers
Office of the U.S. Trade
Representative
Executive Office of the President

Francis J. Turpin
Director
Office of International
    Harmonization
National Highway Traffic Safety
    Administration
U.S. Department of Transportation

Debra H. van Opstal
Fellow in Science and Technology
Political-Military Studies Program
Center for Strategic and International
    Studies

Jerrold L. Wagener
Senior Technical Consultant
Amoco Production Company

Caroline Wagner
Senior Analyst
Rand Corporation

William G. Wagner
Technical Standards Division
    Manager
Society of Automotive Engineers

David Walters
Acting Assistant U.S. Trade
    Representative for Economic
    Affairs and Chief Economist
Office of the U.S. Trade
    Representative
Executive Office of the President

Les Weinstein
Attorney
U.S. Food and Drug Administration

Martin Weiss
Assistant Professor
University of Pittsburgh

Fritz Whittington
Texas Instruments

Fred H. Williamson
Director, Imaging Technology Policy
Eastman Kodak Company

George T. Willingmyre
Vice President
Washington Operations
American National Standards Institute

Mel Woinsky
Senior Manager, Technical Industry
    Standards
Northern Telecom Inc.

Dorothy Zolandz
National Institute of Standards and
    Technology

Joseph F. Zimmer
Deputy Associate Administrator
Office of Management and Budget

Joseph S. Zajaczkowski
Staff Consultant
Standards Coordination
Storage Technology Corp.

**International Standards,
Conformity Assessment, and U.S.
Trade Policy Project Committee**
(attending)

Gary Hufbauer, *Chairman*
Senior Fellow
Institute for International Economics

Stanley Besen
Vice President
Charles River Associates

Dennis Chamot
Associate Executive Director
Commission on Engineering and
    Technical Systems
National Research Council

Leonard Frier
President
MET Laboratories

Steven R. Hix
Chairman
In Focus Systems

Ivor N. Knight
Vice President, Business Technology
    and Standards
COMSAT World Systems

Gerald H. Ritterbusch
Manager of Product Safety and
    Environmental Control
Caterpillar, Inc.

Richard J. Schulte
Senior Vice President, Laboratories
American Gas Association

Lawrence Wills
IBM Director of Standards
IBM Corporation

Academy Complex

Stephen A. Merrill
Executive Director
Board on Science, Technology, and
    Economic Policy and
Director, Academy Industry Program

Lois Perrolle
Associate Director
Academy Industry Program

Daniel LaRue Gross
Staff Associate
Academy Industry Program

Shirley Cole
Administrative Assistant
Academy Industry Program

John S. Wilson
Project Director
International Standards, Conformity
    Assessment, and U.S. Trade
    Policy Project
Board on Science, Technology and
    Economic Policy
National Research Council

John M. Godfrey
Research Associate
International Standards, Conformity
    Assessment, and U.S. Trade
    Policy Project
Board on Science, Technology and
    Economic Policy
National Research Council

Patrick P. Sevcik
Project Assistant
International Standards, Conformity
    Assessment, and U.S. Trade
    Policy Project
Board on Science, Technology and
    Economic Policy
National Research Council

B

Legislative Request for the Study: Public Law 102-245

Public Law 102-245, February 14, 1992

"American Technology Preeminence Act of 1991"
Title V — Studies and Reports
Sec. 508. Study of Testing and Certification

(a) Contract with National Research Council. —Within 90 days after the date of enactment of this Act and within available appropriations, the Secretary shall enter into a contract with the National Research Council for a thorough review of international product testing and certification issues. The National Research Council will be asked to address the following issues and make recommendations as appropriate:

(1) The impact on United States manufacturers, testing and certification laboratories, certification organizations, and other affected bodies of the European Community's plans for testing and certification of regulated and nonregulated products of non-European origin.

(2) Ways for United States manufacturers to gain acceptance of their products in the European Community and in other foreign countries and regions.

(3) The feasibility and consequences of having mutual recognition agreements between testing and certification organizations in the United States

and those of major trading partners on the accreditation of testing and certification laboratories and on quality control requirements.

(4) Information coordination regarding product acceptance and conformity assessment mechanisms between the United States and foreign governments.

(5) The appropriate Federal, State, and private roles in coordination and oversight of testing, certification, accreditation, and quality control to support national and international trade.

(b) Membership. —In selecting the members of the review panel, the National Research Council shall consult with and draw from, among others, laboratory accreditation organizations, Federal and State government agencies involved in testing and certification, professional societies, trade associations, small business, and labor organizations.

(c) Report. —A report based on the findings and recommendations of the review panel shall be submitted to the Secretary, the President, and Congress within 18 months after the Secretary signs the contract with the National Research Council.

APPENDIX

C

Biographical Information on Committee and Staff

COMMITTEE MEMBERS

GARY C. HUFBAUER, *Chairman*, is the Reginald Jones Senior Fellow at the Institute for International Economics (IIE). Prior to his position with the IIE, he was the Marcus Wallenberg Professor of International Financial Diplomacy at Georgetown University. From 1977 to 1980, he served as Deputy Assistant Secretary in the U.S. Treasury where he was responsible for trade and investment policy during the Tokyo Round of the General Agreement on Tariffs and Trade (GATT). Previously, he was Director of the International Tax Staff at the U.S. Treasury. He has published numerous books and articles on international trade, finance, and tax policy.

DENNIS CHAMOT is an Associate Executive Director with the Commission on Engineering and Technical Systems of the National Research Council. From 1969 until 1973 he was with E.I. DuPont de Nemours as a research chemist. In 1974 he became Assistant to the Executive Secretary, Council of AFL-CIO Unions for Professional Employees. The AFL-CIO's Department for Professional Employees was chartered in 1977, at which time he was appointed Assistant Director. He became Associate Director in 1984. He was appointed Executive Assistant to the President in 1990. He has served as an active member of numerous panels and committees; currently serves as a member of the Advisory Committee of the Council on Competitiveness' study of a strategic assessment of national technological priorities; and is a member of the National Research Council's Committee on Review of IRS Information Systems Modernization.

LEONARD FRIER is President of MET Laboratories, Inc., the first licensed Nationally Recognized Testing Laboratory in the United States. He is the founder and former President of MET Electrical Testing Company, Inc. He is a founding and charter member of the National Electrical Testing Association and member of American National Standards Institute (ANSI), American Society of Testing and Materials (ASTM), American Council of Independent Laboratories, and Telephone Industry Association. He is also Chairman of the American National Standards Institute's Z34 Committee of Certification and member of the ANSI Certification Committee.

STEVEN R. HIX is chairman and CEO of Sarif, Inc., Vancouver, Washington. Prior to founding the company, Hix was a cofounder of In Focus Systems, a manufacturer of flat panel displays. Hix was a cofounder of Planar Systems, Inc., where he held senior management positions from 1983 to 1986. Prior to 1983, Hix held management positions at Sigma Research, Inc., Tektronix, Inc., and Watkins Johnson Corporation. He also served for 21 years with the United States Navy, including 10 years as a project design engineer for the Joint Chiefs of Staff.

IVOR N. KNIGHT is President, Knight Communications Consultants, Clarksburg, Maryland. Prior to 1994, he was Vice President, Business Technology and Standards, for COMSAT World Systems, where he was involved with a broad range of communications satellite technology development and business activities. He is a founding member of the ANSI Accredited Standards Committee T1—Telecommunications, and the immediate past Chairman of that organization. He is a member of the Institute of Electrical and Electronics Engineers' Standards Board, a Trustee of the Pacific Telecommunications Council, and the Chairman of the International Telecommunications Union Intersector Coordination Group on all Satellite Matters. He also serves in an advisory capacity to government and industry on international meetings and conferences dealing with telecommunications, trade, and technology.

DAVID C. MOWERY is Associate Professor of Business and Public Policy in the Walter A. Haas School of Business at the University of California, Berkeley. He has also taught at Carnegie-Mellon University. His research deals with the economics of technological innovation and with the effects of public policies on innovation. He has served on several National Research Council panels. In 1988, he served in the Office of the United States Trade Representative as a Council on Foreign Relations International Affairs Fellow. He has also testified before congressional committees and has consulted with various federal agencies and industrial firms. His publications include *Technology and the Pursuit of Economic Growth*; *Alliance Politics and Economics: Multinational Joint Ventures in Commercial Aircraft*; *Technology and Employment: Innovation and Growth in the U.S. Economy*; *The Impact of Technology Change on Employment* *and Economic Growth*; and *International Collaborative Ventures in U.S. Manufacturing.*

MICHAEL M. O'MARA is Business Leader of GE Plastics' Cycolac Business and former director of research and development at General Electric's (GE's) Corporate Research facility in Schenectady, New York. His activities range from research on materials such as diamonds, composites, and new polymers to bioremediation of waste to medical imaging. He was employed by the chemical division of BF Goodrich from 1968 until 1988. He resigned as a senior vice-president of Goodrich's Geon Vinyl Division to join GE. While at Goodrich, he was a manager of the thermoplastic polyurethane business from 1978 to 1981, and vice-president of research and development of the chemical division from 1981 to 1984. In addition to his membership in ASTM, he is a member of the American Chemical Society, the American Association for the Advancement of Science, the advisory board of the *Journal of Applied Polymer Science*, and the Expert Panel of the Hazards of MSW Recycling (Plastics). He was a former member of the Board of Assessment of the National Institute of Standards and Technology (NIST) Programs and formerly chairman of the following: Society of the Plastics Industry's (SPI) Coordinating Committee on Fire Safety; Oversight Panel, Center for Fire Research (NIST); and the first chairman of the Technical Committee of the Vinyl Institute (SPI).

GERALD H. RITTERBUSCH is Manager of Product Safety and Environmental Control for Caterpillar Inc., Peoria, Ill. He is responsible for the product safety and environmental regulations involvement of Caterpillar staff with governments worldwide. He serves in a number of standards development organizations such as the Society of Automotive Engineers (SAE), and American Society of Mechanical Engineers and, as Chairman of the SAE Technical Standards Board International Harmonization Committee and Chairman of the SAE Performance Review Board. He is vice-chairman of the ANSI Company Member Council Executive Committee, Chairman of the International Standards Organization Technical Committee 127—Earthmoving Machinery. He is also active in the industry trade association, previously serving as Chairman of the Technical Council of the Equipment Manufacturers Institute.

RICHARD J. SCHULTE was appointed Senior Vice President, Laboratories of the American Gas Association (AGA) in 1992. In this position he has direct responsibility for the successful operation of AGA gas appliance certification and research programs and for administrative support of three national standards committees accredited by the American National Standards Institute. Since 1985, he has served on the board of directors of ANSI. He is currently chairman of an ANSI panel studying improvements in the voluntary standardization procedures used by all U.S. industry. In August 1990, he was honored by the Association of Home Appliance Manufacturers for his outstanding efforts to harmonize gas appliance standards on a global basis. He was named AGA's Marketing Executive of the Year in 1991 for his achievements and contributions to AGA's product certification and research programs.

SUSAN C. SCHWAB is Director of Motorola, Inc.'s Corporate Business Development Office at their corporate headquarters in Schaumburg, Illinois. Prior to joining Motorola she was Assistant Secretary and Director General of the United States and Foreign Commercial Service at the Department of Commerce (DoC) in Washington, D.C. from 1989 to 1993. Prior to going to the DoC she was Legislative Assistant, Chief Economist, and Legislative Director to Senator John Danforth from 1981 to 1989. She previously held the position of trade policy officer at the U.S. Embassy in Tokyo and, prior to that, was an international economist at the Office of the U.S. Trade Representative.

MICHAEL B. SMITH is President, SJS Advanced Strategies, Washington, D.C. From 1983 to 1988 he was Senior Deputy U.S. Trade Representative with rank of Ambassador at the Office of the United States Trade Representative. Prior to that position he was U.S. Ambassador to the GATT and Deputy United States Trade Representative from 1979 to 1983. In previous positions he was Chief Textile Negotiator of the United States with rank of Ambassador, Office of the United States Trade Representative (1975-1979); Chief of Fibers and Textiles Division, Economic Bureau, U.S. Department of State (1974-1975); Deputy Chief of Fibers and Textiles Division (1973-1974); and Staff Assistant to the President of the United States (1971-1973); and has held many U.S. Foreign Service positions. He is a founding member of the Pacific Economic Cooperation Council and member of the Atlantic Council.

LAWRENCE L. WILLS is IBM Director of Standards with the IBM Corporation. At IBM he is responsible for programs covering products and operations related to standards, and he directs IBM's participation in and relations with external standards organizations throughout the world. Mr. Wills serves on the U.S. Department of State Advisory Committee on International Communications and Information Policy, the Board of the U.S. Telecommunications Training Institute, and the Board of Directors of the Information Industry Association. He is Chairman of ANSI's Board of Directors and is Chairman of the International Advisory Committee and the Board of Trustees of the International Fund.

PROFESSIONAL STAFF

JOHN S. WILSON is Project Director and Senior Staff Officer at the National Academies of Sciences and Engineering and National Research Council. He is also a Visiting Fellow at the Institute for International Econmics (IIE) in 1995. His duties at the NRC include serving as a senior staff officer for the

Policy Division, where he is responsible for research and analysis of U.S. and international economic and technology policies. From 1989 to 1992, Mr. Wilson served as Project Director for the National Academies of Sciences and Engineering study of the government's role in civilian technology. Mr. Wilson has also held positions with the congressional Office of Technology Assessment. He was on the staff of the Public Affairs Division of Pfizer, Inc., served as Assistant to the President on the Committee for Economic Development, and Adjunct Professor of International Affairs at Georgetown University (1993-1994). He contributed to the work of the President's Commission on Setting a National Agenda for the Eighties and the President's Commission on Industrial Competitiveness. He is the author of numerous papers and reports on technology, trade, and economic policy issues.

JOHN M. GODFREY is Research Associate with the National Academy of Sciences and the National Research Council. Before joining the staff of the Academy in 1993, Mr. Godfrey was a science policy analyst in the Arlington, Virginia offices of SRI International, a not-for-profit research institute based in Menlo Park, California. His research interests and activities at SRI included comparative international science and technology policy, quantitative analysis of R&D expenditures and manpower, and assessment of programs for international scientific cooperation. His publications as author or coauthor include *Institutional Linkages Between U.S. and Foreign Universities and Research Centers: A Report to the National Science Foundation*, "NAFTA and the Mexican Energy Sector," and *R&D Expenditures of Selected Industrialized Nations* (report and data on diskette to the National Science Foundation).

PATRICK P. SEVCIK is Project Assistant with the Board on Science, Technology, and Economic Policy (STEP) of the National Academy of Sciences and National Research Council (NRC). He works on several projects as an assistant to John Wilson, Senior Staff Officer at STEP, one of which is the International Standards, Conformity Assessment, and U.S. Trade Policy Project overseen by the STEP Board. Prior to his work at the NRC, Mr. Sevcik was an Assistant Program Officer with the International Republican Institute from 1990 to 1993. In this capacity he specialized in working with democratically aligned groups in Central and Eastern Europe and assisting these groups in the promotion of democratic ideals and free and fair elections. He has worked extensively in Bosnia-Herzegovina, Macedonia, Serbia, Montenegro, Bulgaria, Romania, Hungary, the Czech Republic, and the Republic of Slovakia. He has also held positions at the White House in the Office of Political Affairs (1989-1990) and on Capitol Hill (1987-1988) in the Office of Representative John DioGuardi (R-NY).

APPENDIX
D

Glossary and Acronyms

GLOSSARY

Standard   A prescribed set of conditions or requirements concerning definitions of terms; specification of performance, operation, or construction; delineation of procedures; or measurement of quantity and/or quality in describing features of products, processes, systems, interfaces, or materials.

Compatibility Standard   A standard that defines the interface between components of a system or network, allowing them to function or communicate with one another.

De Facto Standard   A standard arising from uncoordinated processes in the competitive marketplace. When a particular set of product or process specifications gains market share such that it acquires authority or influence, that set of specifications is then considered a de facto standard.

Design Standard   A standard that defines a product's characteristics in terms of how it is to be constructed.

Mandatory Standard

A standard set by government. A procurement standard specifies requirements that must be met by suppliers to government. A regulatory standard may set safety, health, environmental, or related criteria. Voluntary standards developed for private use often become mandatory when referenced within government regulation or procurement.

Performance Standard

A standard that defines a product's characteristics in terms of how it is to function. Because this type of standard leaves open to the designer the issue of how the product achieves the desired functionality, performance standards are considered less restrictive of innovation than design standards.

Voluntary Consensus Standard

A standard arising from a formal, coordinated process in which key participants in a market seek consensus. Use of the resulting standard is voluntary. Key participants may include not only designers and producers, but also consumers, corporate and government purchasing officials, and regulatory authorities.

CONFORMITY ASSESSMENT

Conformity Assessment

The determination of whether a product or process conforms to particular standards or specifications. Activities associated with conformity assessment may include testing, certification, accreditation, and quality assurance system registration.

Accreditation

The process of evaluating testing facilities for competence to perform specific tests using standard test methods. The test methods for which a facility seeks accreditation may or may not be associated with a particular certification program.

Certification

The process of providing assurance that a product or service conforms to one or more standards or specifications. Some, but not all, certification programs require that an accredited laboratory perform any required testing.

Certification Mark   A sign or symbol that is used exclusively by the operator of a third-party certification program to identify products or services as being certified. In the United States, certification marks are registered with the U.S. Patent Office. A well-known example is the Underwriters Laboratories' "UL" mark.

Manufacturer's Declaration of Conformity   A form of certification in which a manufacturer or supplier declares that his or her product meets one or more standards based on (1) confidence in the manufacturer's own quality assurance program, and/or (2) the results of testing the manufacturer performs or has performed on his or her behalf.

Mutual Recognition Agreement   Agreement between conformity assessment entities to accept some or all aspects of one another's work. Both public- and private-sector parties may be involved.

New Approach Directives   Directives produced by the Commission of the European Union specifying "essential requirements" for safety that regulated goods must meet in order to be placed on the market anywhere within the European Union. Producers may demonstrate conformity to the directives via a choice among eight Commission-specified assessment modules, three of which involve quality system registration to portions of the International Organization for Standardization (ISO) 9000 series.

As of 1994, new approach directives exist for toys, simple pressure vessels, construction products, electromagnetic compatibility, gas appliances, personal protective equipment, machinery, weighing instruments, certain medical devices, and telecommunications terminal equipment.

Recognition   The process of evaluation and designation, by a government entity, that an accreditation program is competent to carry out its activities.

Third-Party Certification   A form of certification in which the producer's claim of conformity is validated in a third-party certification program by a competent body other than one controlled by the producer.

Quality Assurance The procedures and resources a manufacturer uses to control variables in product design, production, and handling in order to produce a product of consistent quality that meets defined specifications or standards.

Environmental Management System The procedures and resources an organization uses to implement its policies regarding the impact of its activities, products, and services on the environment.

ISO 9000 Standard Series A series of five international standards for quality assurance management systems (ISO 9000, 9001, 9002, 9003, and 9004) published in 1987 by the International Organization for Standardization and revised in 1994. The standards are being applied worldwide in an increasingly broad range of circumstances, including government regulations and public- and private-sector purchasing specifications.

Quality System Registration Assessment and periodic audit of the adequacy of a producer's quality assurance system by a third party known as a quality system registrar. ISO 9000 quality system registration, for example, certifies that a producer's system conforms to the registrar's interpretation of one or more of the ISO 9000 standards; interpretations may vary among registrars.

ACRONYMS

A2LA	American Association for Laboratory Accreditation
AAMI	Association for the Advancement of Medical Instrumentation
ACIL	American Council of Independent Laboratories
AFNOR	Association Francaise de Normalisation (French Standardization Association)
AGA	American Gas Association
ANSI	American National Standards Institute
APEC	Asia Pacific Economic Cooperation Council
ASC-T1	Accredited Standards Committee T1—Telecommunications
ASME	American Society of Mechanical Engineers
ASQC	American Society for Quality Control
ASTM	American Society for Testing and Materials

ATIS	Alliance for Telecommunications Industry Solutions (formerly, the Exchange Carriers Standards Association, ECSA)
ATP	Advanced Technology Program, NIST
BSI	British Standards Institution
CASCO	ISO Council Committee on Conformity Assessment
CBEMA	Computer and Business Equipment Manufacturers Association
CCITT	ITU International Telephone and Telegraph Consultative Committee (recently reorganized as the ITU Telecommunication Standardization Sector, ITU-T)
CEN	European Committee for Standardization
CENELEC	European Committee for Electrotechnical Standardization
CNAEL	Committee on National Accreditation of Environmental Laboratories
COPANT	Pan American Standards Commission
COS	Corporation for Open Systems
CPSC	U.S. Consumer Product Safety Commission
CSA	Canadian Standards Association
DIN	Deutsches Institut fur Normung (German Institute for Standardization)
DoC	U.S. Department of Commerce
DoD	U.S. Department of Defense
DSN	Dewan Standardisasi Nasional (National Standardization Council of Indonesia)
EFTA	European Free Trade Association
EOTC	European Organization for Testing and Certification
EPA	U.S. Environmental Protection Agency
ETSI	European Telecommunications Standards Institute
EU	European Union (formerly, the European Community)
FCC	Federal Communications Commission
FDA	U.S. Food and Drug Administration
FY	fiscal year
GAMA	Gas Appliance Manufacturers Association
GATT	General Agreement on Tariffs and Trade
GDP	gross domestic product
GSA	U.S. General Services Administration
ICSP	Interagency Committee on Standards Policy

IEC	International Electrotechnical Commission
IEEE	Institute of Electrical and Electronics Engineers
IFAC	Industry Functional Advisory Committee, U.S. Department of Commerce
IIE	Institute for International Economics
IPR	intellectual property rights
ISAC	Industry Sectoral Advisory Committee, U.S. Department of Commerce
ISO	International Organization for Standardization
IT	information technology
ITA	International Trade Administration, U.S. Department of Commerce
ITU	International Telecommunication Union, United Nations
ITU-T	ITU Telecommunication Standardization Sector
JISC	Japan Industrial Standards Committee
JSA	Japan Standards Association
JTC1	ISO/IEC Joint Technical Committee 1—Information Technology
MAP	Manufacturing Automation Protocol
MBNQA	Malcolm Baldrige National Quality Award
MITI	Japanese Ministry for International Trade and Industry
MOU	memorandum of understanding
MRA	mutual recognition agreement
NADCAP	National Aerospace and Defense Contractors Accreditation Program
NAFTA	North American Free Trade Agreement
NAM	National Association of Manufacturers
NAVFAC	Naval Facilities Engineering Command
NCASR	National Conformity Assessment System Recognition Program (proposed)
NCSCI	National Center for Standards and Certification Information, NIST
NCWM	National Conference on Weights and Measures
NEMA	National Electrical Manufacturers Association
NFPA	National Fire Protection Association
NHTSA	National Highway and Traffic Safety Administration
NIST	National Institute of Standards and Technology, U.S. Department of Commerce
NRTL	Nationally Recognized Testing Laboratory Program, OSHA
NSSN	National Standards Systems Network
NTB	nontariff trade barrier

NTE	National Trade Estimate
NVCASE	National Voluntary Conformity Assessment Systems Evaluation Program, NIST
NVLAP	National Voluntary Laboratory Accreditation Program, NIST
OMB	Office of Management and Budget, Executive Office of the President, U.S.
OSHA	Occupational Safety and Health Administration
OSI	Open Systems Interconnection
PASC	Pacific Area Standards Congress
PAT	Process Action Team on Military Specifications and Standards, U.S. Department of Defense
QML	Qualified Manufacturers List
QPL	Qualified Product List
R&D	research and development
RAB	Registrar Accreditation Board
RVC	Raad voor de Certificatie (Dutch Council for Certification)
SAE	Society of Automotive Engineers
SASO	Saudi Arabian Standards Organization
SCC	Standards Council of Canada
SCI	Standards Code and Information Program, NIST
SDO	standards developing organization
SPS	sanitary and phytosanitary
TAG	technical advisory group
TBT	technical barrier to trade
TC	technical committee
TQM	total quality management
UL	Underwriters Laboratories, Inc.
US&FCS	U.S. and Foreign Commercial Service, U.S. Department of Commerce
USDA	U.S. Department of Agriculture
USNC	U.S. National Committee of the IEC
USNRC	U.S. Nuclear Regulatory Commission
USTR	United States Trade Representative, Executive Office of the President
WTO	World Trade Organization

APPENDIX
E

Selected Bibliography

ACIL: The Association of Independent Scientific, Engineering and Testing Firms. *ACIL Newsletter* (June 1994).

ACIL: The Association of Independent Scientific, Engineering and Testing Firms. *Directory: A Guide to Leading Independent Testing, Research and Inspection Firms in America.* 1992-1993 edition. Washington, D.C.: ACIL, 1992.

American Association for Laboratory Accreditation. *A2LA 1994 Directory of Accredited Laboratories.* Gaithersburg, Md.: A2LA, 1994.

American Association for Laboratory Accreditation and Standards Council of Canada. *Mutual Recognition Agreement between the Standards Council of Canada (SCC) and the American Association for Laboratory Accreditation (A2LA).* Signed June 23, 1994.

American Association for Laboratory Accreditation. *A2LA 1993 Annual Report.* Gaithersburg, Md.: A2LA, 1993.

American Council of Independent Laboratories. *ACIL 1993 Annual Report.* Washington, D.C.: ACIL, 1993.

American Gas Association. *American Gas Association Laboratories Information Booklet.* Cleveland, Oh.: AGAL, 1994.

American National Standards Institute. *American National Standards for Certification Z34.1.* New York: ANSI, 1994.

American National Standards Institute. *1993 Annual Report.* New York: ANSI, 1993.

American National Standards Institute. *The U.S. Voluntary Standardization System: Meeting the Global Challenge.* 2nd edition. New York: ANSI, 1993.

American National Standards Institute. *Guidelines for Implementation of The ANSI Patent Policy.* New York: ANSI, 1992.

American National Standards Institute. *1992 Annual Report.* New York: ANSI, 1992.

American National Standards Institute. *A Cooperative Standards-Setting Process will Contribute to American Competitiveness and Economic Health.* Position paper by the American National Standards Institute. New York: ANSI, December 1992.

American National Standards Institute. *ANSI Analysis of National Institute of Standards and Tech-*

213

nology, Office of Standards Service, Hearing on: Improving U.S. Participation in International Standards Activities. April 3-5, 1990. New York: ANSI, 1990.

American Society of Mechanical Engineers. "Annual Report for 1992/1993", In *Mechanical Engineering* 115, AR-11 (November 1993).

American Society of Mechanical Engineers. *1992 ASME Boiler & Pressure Vessel Code.* Fairfield, N.J.: ASME, 1992.

American Society of Mechanical Engineers. *The Why and How of Codes & Standards.* Fairfield, N.J.: ASME, 1992.

American Society for Testing and Materials. *Internal Memorandum,* May 11, 1994. Philadelphia, Pa.: ASTM, 1994.

American Society for Testing and Materials. *ASTM 1993 Annual Report.* Philadelphia, Pa.: ASTM, 1993.

American Society for Testing and Materials. *ASTM 1992 Annual Report.* Philadelphia, Pa.: ASTM, 1992.

American Society for Testing and Materials. *ASTM and Voluntary Consensus Standards.* Philadelphia, Pa.: ASTM, n.d.

Asia-Pacific Economic Cooperation. Secretariat. *Achieving the APEC Vision; Free and Open Trade in the Asia Pacific.* Second report of the Eminent Persons Group, August 1994.

Asia-Pacific Economic Cooperation. Experts Meeting on Standards and Conformance. Committee on Trade and Investment. *Closer Cooperation with International Standards and Conformance Bodies: Submitted by New Zealand.* Sixth Ministerial Meeting. Document III.6, 1994.

Asia-Pacific Economic Cooperation. Experts Meeting on Standards and Conformance. Committee on Trade and Investment. *Proposed Declaration on an APEC Standards and Conformance Framework: Submitted by Australia.* Sixth Ministerial Meeting. Document III.6.5, 1994.

Asia-Pacific Economic Cooperation. Experts Meeting on Standards and Conformance. Committee on Trade and Investment. *Report of the APEC Formal Meeting on Standards and Conformance: Bali, 13 May 1994.* Sixth Ministerial Meeting. Document III.6.6 (Final), 1994.

Association Francaise de Normalisation. *Information File on: The NF-Environment Mark and the European Ecolabel in France.* Paris: Association Francaise de Normalisation, August 1993.

Baldwin, Robert E. "The Economics of the GATT." In Peter Oppenheimer, ed. *Issues in International Economics.* Oxford International Symposia (5)82-93. London: Oriel Press, 1980. In Patrick Low. *Trading Free: The GATT and U.S. Trade Policy.* New York: The Twentieth Century Fund, 1993.

Baldwin, Robert E. *Nontariff Distortions of International Trade.* Washington, D.C.: Brookings Institution, 1970.

Barrier, Michael, and Amy Zuckerman. "Quality Standards the World Agrees On." In *Nation's Business* (May 1994): 71-73.

Bayard, Thomas O., and Kimberly Ann Elliott. *Reciprocity and Retaliation in U.S. Trade Policy.* Washington, D.C.: Institute for International Economics, 1994.

Bello, Judith H., and Alan F. Holmer. "The GATT Uruguay Round: Its Significance for U.S. Bilateral Trade with Korea and Taiwan." In *Michigan Journal of International Law* 11, no. 2 (1990): 307-325.

Bello, Judith Hippler, and Alan F. Homer. "The Heart of the 1988 Trade Act: A Legislative History of Amendments to Section 301." In *Stanford Journal of International Law* (Fall 1988): 1-44. In Patrick Low. *Trading Free: The GATT and U.S. Trade Policy.* New York: The Twentieth Century Fund, 1993.

Berg, Sanford V., and Dennis L. Weisman. "A Guide to Cross-Subsidization and Price Predation." In *Telecommunications Policy* (August 1992): 447-459. Butterworth-Heinemann Ltd.

Berg, Sanford V. "The Production of Compatibility: Technical Standards as Collective Goods." In *Kyklos* 42 (1989): 361-383.

Berg, Sanford V. "Technical Standards as Public Goods: Demand Incentives for Cooperation Behavior." In *Public Finance Quarterly* 17, no. 1 (1989): 29-54.

Berg, Sanford V. "Public Policy and Corporate Strategies in the AM Stereo Market." In *Product Standardization and Competitive Strategy*, H. Landis Gabel, ed. Amsterdam: Elsevier Science Publishers B.V., 1987.

Besen, Stanley M., and Joseph Farrell. "Choosing How to Compete: Strategies and Tactics in Standardization." In *Journal of Economic Perspectives* 8, no. 2 (Spring 1994): 1-15.

Besen, Stanley M. "The Standards Processes in Telecommunications and Information Technology." Charles River Associates. Preliminary draft of paper presented at the *SPRU–OECD International Workshop on Standards, Innovation, Competitiveness and Policy*. University of Sussex. November 10-12, 1993.

Besen, Stanley M. *Telecommunications and Information Technology Standard-Setting in Japan: A Preliminary Survey*. A RAND Note, N-3204-CUSJR. Santa Monica, Calif.: RAND, 1991.

Besen, Stanley M., and Joseph Farrell. "The Role of the ITU in Standardization: Pre-eminence, impotence or rubber stamp?" RAND. RP-100. Reprinted from *Telecommunications Policy* (August 1991): 311-321.

Besen, Stanley M. "The European Telecommunications Standards Institute: A preliminary analysis." In *Telecommunications Policy* (1990): 521-530.

Besen, Stanley M., and Garth Saloner. "The Economics of Telecommunications Standards." In *Changing the Rules: Technological Change, International Competition, and Regulation in Communications*, Robert W. Crandall and Kenneth Flamm, eds. Pp. 177-220. Washington, D.C.: Brookings Institution, 1989.

Besen, Stanley M., and Leland L. Johnson. *Compatability Standards, Competition, and Innovation in the Broadcasting Industry*. Prepared for the National Science Foundation. R-3453-NSF. Washington, D.C.: RAND, 1986.

Betancourt, Diego. "Strategic Standardization Management: A Strategic Macroprocess Approach to the New Paradigm in the Competitive Business Use of Standardization." Preliminary research paper prepared for the *ANSI Company Member Council-Executive Committee*. Cambridge, Mass.: Polaroid Corporation, 1993.

Bhagwati, Jagdish. "Beyond NAFTA: Clinton's Trading Choices." In *Foreign Policy* (Summer 1993): 155-162.

Bhagwati, Jagdish N. "Trade and the Environment." In *The American Enterprise* (May/June 1993): 44-49.

Bhagwati, Jagdish N., and Hugh T. Patrick, eds. *Aggressive Unilateralism: America's 301 Trade Policy and the World Trading System*. Ann Arbor, Mich.: University of Michigan Press, 1990. In Patrick Low. *Trading Free: The GATT and U.S. Trade Policy*. New York: The Twentieth Century Fund, 1993.

Bhagwati, Jagdish. "The Threats to the World Trading System." In *The World Economy* 15, no. 4 (1992): 443-456.

Block, Marilyn R. "ISO/TC 207: Developing an International Environmental Management Standard." In *The European Marketing Guide: Economic, Environmental, Legal & Social Strategies*. Part of The Complete European Digest II, no. 3 (March 1994). Atlanta, Ga.: SIMCOM, 1994.

Boerner, Christopher, and Kenneth Chilton. *Recycling's Demand Side: Lesson's from Germany's "Green Dot."* Contemporary Issues Series 59. Center for the Study of American Business, Washington University. St. Louis, Mo.: Washington University, 1993.

Boltuck, Richard, and Robert E. Litan, eds. *Down in the Dumps: Administration of the Unfair Trade Laws*. Washington, D.C.: Brookings Institution, 1991. In Patrick Low. *Trading Free: The GATT and U.S. Trade Policy*. New York: The Twentieth Century Fund, 1993.

Breitenberg, Maureen. *More Questions and Answers on the ISO 9000 Standard Series and Related Issues*. NISTIR 5122. Prepared for the National Institute of Standards and Technology, U.S. Department of Commerce. Gaithersburg, Md.: NIST, 1993.

Breitenberg, Maureen. *Laboratory Accreditation in the United States*. NISTIR 4576. Prepared for the National Institute of Standards and Technology, U.S. Department of Commerce. Gaithersburg, Md.: NIST, 1991.

Breitenberg, Maureen, ed. *Directory of Federal Laboratory Accreditation/Designation Programs*. NIST Special Publication 808. National Institute of Standards and Technology, U.S. Department of Commerce. Washington, D.C.: U.S. Government Printing Office, 1991.

Breitenberg, Maureen, ed. *Directory of U.S. Private-Sector Product Certification Programs*. NIST Special Publication 774. National Institute of Standards and Technology, U.S. Department of Commerce. Washington, D.C.: U.S. Government Printing Office, 1989.

Breitenberg, Maureen, ed. *Directory of International and Regional Organizations Conducting Standards-Related Activities*. NIST Special Publication 767. National Institute of Standards and Technology, U.S. Department of Commerce. Gaithersburg, Md.: NIST, 1989.

Breitenberg, Maureen. *The ABC's of Certification Activities in the United States*. NBSIR 88-3821. Prepared for the National Bureau of Standards, U.S. Department of Commerce. Gaithersburg, Md.: NBS, 1988.

Breitenberg, Maureen, ed. *Directory of Federal Government Certification Programs*. NBS Special Publication 739. National Bureau of Standards, U.S. Department of Commerce. Washington, D.C.: U.S. Government Printing Office, 1988.

Breitenberg, Maureen, ed. *Index of Products Regulated By Each State*. NBSIR 87-3608. Prepared for the National Bureau of Standards, U.S. Department of Commerce. Gaithersburg, Md.: NBS, 1987.

Breitenberg, Maureen. *The ABC's of Standards-Related Activities in the United States*. NBSIR 87-3576. Prepared for the National Bureau of Standards, U.S. Department of Commerce. Gaithersburg, Md.: NBS, 1987.

Brock, William, and Robert Hormats, eds. *The Global Economy: America's Role in the Decade Ahead*. New York: W.W. Norton, 1990. In Patrick Low. *Trading Free: The GATT and U.S. Trade Policy*. New York: The Twentieth Century Fund, 1993.

Brown, Ronald H. *Toward a National Export Strategy: U.S. Exports = U.S. Jobs*. Report to the United States Congress. Washington, D.C.: Trade Promotion Coordinating Committee, 1993.

Cadwalader, Wickersham & Taft. *The Antitrust Implications of Voluntary Standards Activities—An Update*. Memorandum to the Board of Directors of the American National Standards Institute, November 18, 1988.

Cargill, Carl F. *Information Technology Standardization: Theory, Process, and Organizations*. Bedford, Mass.: Digital Press, 1989.

Cargill, Carl, and Martin Weiss, "Consortia in the Standards Development Process," In *Journal of the American Society for Information Science* 43, no. 8 (1992): 559-565

Cascio, Joe. Chair, U.S. Technical Advisory Group to ISO Technical Committee 207, presentation at 1994 ANSI Annual Public Conference, Washington, D.C., March 4, 1994.

Cassell, Jordan W., and Robert L. Crosslin. *Benefits of the Defense Standardization Program*. Bethesda, Md.: Logistics Management Institute, 1991.

Cave, Martin E., Paul A. David, and Mark Shurmer. *The Political Economy of International Standards Institutions: Causes and Consequences of Changes in Standards-Setting Regimes for Information and Communications Technology*. Paper submitted to the ERSC Programme on Global Economic Institutions, 1993.

CEEM Information Services. "Big Three Standards to be Rolled Out This Month," In *Quality Systems Update* (July, 1994).

Center for Economic Policy Research. *New Trade Theories: A Look at the Empirical Evidence*. A CEPR Conference Report, May 1993 in Milan, Italy. Great Britain: CEPR, 1994.

Center for Strategic and International Studies, Working Group on Military Specifications and Standards. *Road Map for Milspec Reform.* Washington, D.C.: CSIS, 1993.

Cheit, Ross E. *Setting Safety Standards: Regulation in the Public and Private Sectors.* Berkeley, Calif.: University of California Press, 1990.

Cline, William R. "'Reciprocity': A New World Approach to World Trade Policy?" Washington, D.C.: Institute for International Economics, 1982. In Patrick Low. *Trading Free: The GATT and U.S. Trade Policy.* New York: The Twentieth Century Fund, 1993.

Clinton, William J. *Report to the Congress on Recommendations on Future Free Trade Area Negotiations.* Washington, D.C.: Executive Office of the President, July 1, 1994.

Coffield, Shirley A. "Section 310 of the Trade Act of 1974: New Life in the Old Dog." In *Federal Bar News and Journal* 33, no. 6 (July/August 1986): 248-251. In Patrick Low. *Trading Free: The GATT and U.S. Trade Policy.* New York: The Twentieth Century Fund, 1993.

Cohen, Robert B., and Kenneth Donow. *Telecommunications Policy, High Definition Television, and U.S. Competitiveness.* Washington, D.C.: Economic Policy Institute, 1989.

Cohen, Robert B., Richard W. Ferguson, and Michael F. Oppenheimer. *Nontariff Barriers to High-Technology Trade.* Boulder, Colo.: Westview Press, 1985.

Commission of the European Communities. *Report on United States Trade and Investment Barriers 1992: Problems of Doing Business with the U.S.* Services of the Commission of the European Communities, 1992.

Computer Systems Policy Project. *Perspectives on the National Information Infrastructure: Ensuring Interoperability.* Washington, D.C.: CSPP, 1994.

Cooke, Patrick W. *An Update of U.S. Participation in International Standards Activities.* NISTIR 89-4124. Prepared for the National Institute of Standards and Technology, U.S. Department of Commerce. Gaithersburg, Md.: NIST, 1989.

Cooke, Patrick W. *A Review of U.S. Participation in International Standards Activities.* NBSIR 88-3698. Prepared for the National Bureau of Standards, U.S. Department of Commerce. Gaithersburg, Md.: NBS, 1988.

Cooney, Stephen. *The Europe of 1992: An American Business Perspective.* Prepared for the National Association of Manufacturers. Washington, D.C.: NAM International Economic Affairs Department, 1992.

Cooper, Richard N. "Industrial Policy and Trade Distortion." In Dominick Salvatore, ed. *The New Protectionist Threat to World Welfare.* Pp. 233-265. New York: North Holland, 1987. In Patrick Low. *Trading Free: The GATT and U.S. Trade Policy.* New York: The Twentieth Century Fund, 1993.

Corrigan, James P. "Is ISO 9000 the Path to TQM?" In *Quality Progress* (May 1994): 33-36.

Crane, Rhonda J. *The Politics of International Standards: France and the Color TV War.* Communication and Information Science series, Melvin J. Voigt, ed. N.J.: Ablex Publishing, 1979.

Curzon, Gerard. *Multilateral Commercial Diplomacy.* London: Michael Joseph, 1965: p. 15-34. In Patrick Low. *Trading Free: The GATT and U.S. Trade Policy.* New York: The Twentieth Century Fund, 1993.

Curzon-Price, Victoria. *The Post-Uruguay Round Trade Outlook: Standards and Technical Barriers to Trade.* Presentation to Conference on New Developments in International Standards and Global Trade. Washington, D.C., March 30, 1994.

Cyert, Richard M., and David C. Mowery, eds. *The Impact of Technological Change on Employment and Economic Growth.* The Committee on Science, Engineering, and Public Policy, National Academy of Sciences. Cambridge, Mass.: Ballinger Publishing Company, 1988.

Dam, Kenneth W. *The GATT: Law and the International Economic Organization.* Chicago: University of Chicago Press, Midway Reprint, 1977: p. 12. In Patrick Low. *Trading Free: The GATT and U.S. Trade Policy.* New York: The Twentieth Century Fund, 1993.

David, Paul A., and Geoffrey S. Rothwell. *Standardization, Diversity, and Learning: A Model for the Nuclear Power Industry.* Stanford, Calif.: Stanford University, 1992-Version.

David, Paul A., and Shane Greenstein. "The Economics of Compatibility Standards: An Introduction to Recent Research." In *Economics of Innovation and New Technology* 1 (1990): 3-41.

David, Paul A., and W. Edward Steinmuller. "The ISDN Bandwagon is Coming, but Who will be there to Climb Aboard?: Quandries in the Economics of Data Communication Networks." In *Economics of Innovation and New Technology* 1 (1990): 43-62.

David, Paul A., and Julie Ann Bunn. "The Economics of Gateway Technologies and Network Evolution: Lessons from Electricity Supply History." Reprinted from *Information Economics and Policy* 3 (1988): 165-202.

David, Paul A. "Some New Standards for the Economics of Standardization in the Information Age." Reprinted from *Economic Policy and Technological Performance*. Partha Dasgupta and Paul Stoneman, eds. Cambridge, U.K.: Cambridge University Press, 1987.

David, Paul A. "Clio and the Economics of QWERTY." In John G. Riley, Wilma St. John, eds. Papers and Proceedings of the Ninety-Seventh Annual Meeting of the American Economic Association. 1984. December 28-30, 1984 in Dallas, Texas. In the *American Economic Review* 75, no. 2 (May 1985): 332-337.

Deardorff, A. V., and R. M. Stern. *Analytical and Negotiating Issues in the Global Trading System*. Mich.: University of Michigan Press, 1994.

Defense Systems Management College. *Standards and Trade in the 1990s: A Source Book for Department of Defense Acquisition and Standardization Management and their Industrial Counterparts*. Washington, D.C.: U.S. Government Printing Office, n.d.

Deloitte & Touche. *ISO 9000 Survey: September 1993 Results*. CEEM Information Services' Quality Systems Update: A Global ISO 9000 Forum and Information Service. Atlanta, Ga.: Deloitte & Touche, 1993.

Destler, I. M. "United States Trade Policy Making in the Uruguay Round." In Henry R. Nau, ed. *Domestic Trade Politics and the Uruguay Round*. New York: Columbia Press, 1989. In Patrick Low. *Trading Free: The GATT and U.S. Trade Policy*. New York: The Twentieth Century Fund, 1993.

Destler, I. M. *American Trade Politics*. 2nd edition. Washington, D.C.: Institute for International Economics, 1992.

Drake-Brockman, Jane, and Kyim Anderson. *The Entwining of Trade Policy and Environmental Issues: Implications for APEC*. Presented at The Second Conference on "APEC: NAFTA/ ASEAN/SAARC." August 29 - September 1, 1994, in Nusa Dua, Bali, Indonesia. Australia: Centre for International Economic Studies, 1994.

Durand, Ian G.; Donald W. Marquardt, Robert W. Peach, and James C. Pyle. "Updating the ISO 9000 Quality Standards: Responding to Marketplace Needs." In *Quality Progress* (July 1993): 23-28.

Eads, George, and Peter Reuter. *Designing Safer Products: Corporate Responses to Product Liability Law and Regulation*. R-3022ICJ. The Institute for Civil Justice. Santa Monica, Calif.: RAND, 1983.

The EC Marketing Guide. *Conformity Assessment: What it is, and What are the Trade Implications for U.S. Exporters? Economic, Environmental, Legal & Social Strategies*. European Community.

Edelman, Peter B. "Japanese Product Standards as Non-Tariff Trade Barriers: When Regulatory Policy becomes a Trade Issue." Stanford University. In *Stanford Journal of International Law* 24, no. 2 (Spring 1988): 389-446.

Electric Power Research Institute. "Comparison of ISO 9000 Requirements to Those of 10 C.F.R. 50 Appendix B: Task Sheet." April 8, 1992.

Elliott, Kimberly Ann, and Thomas Bayard. *Reciprocity and Retaliation: Does Might Make Right in U.S. Trade Policy?* Washington, D.C.: Institute for International Economics. In Patrick Low. *Trading Free: The GATT and U.S. Trade Policy*. New York: The Twentieth Century Fund, 1993.

European Organization for Testing and Certification. *EOTC: Focal Point for Testing & Certification in Europe*. Brochure. October 1993.

European Telecommunications Standards Institute. *Annual Report 1992*. France: ETSI, 1992.

Farrell, Joseph; Hunter K. Monroe, and Garth Saloner. Draft. *Order Statistics, Interface Standards, and Open Systems*. EAG Seminar, January 12, 1994.

Farrell, Joseph. *Choosing the Rules for Formal Standardization*. Draft. Berkeley, Calif.: University of California, 1993.

Farrell, Joseph, and Carl Shapiro. "Standard Setting in High-Definition Television." *Brookings Papers: Microeconomics 1992*. Washington, D.C.: Brookings Institution, 1992.

Farrell, Joseph, and Garth Saloner. "Converters, Compatibility, and the Control of Interfaces." In *The Journal of Industrial Economics* XL, no. 1 (March 1992).

Farrell, Joseph. "The Economics of Standardization: A Guide for Non-Economists." In *An Analysis of the Information Technology Standardization Process*. Elsevier Science Publishers B.V., 1990: pp. 189-198.

Farrell, Joseph. "Standardization and Intellectual Property." American Bar Association. In *Jurimetrics Journal of Law, Science and Technology* 30, no. 1 (Fall 1989): 35-50.

Farrell, Joseph, and Garth Saloner. "Coordination through Committee and Markets." In *RAND Journal of Economics* 19, no. 2 (1988): 235-252.

Farrell, Joseph, and Garth Saloner. "Competition, Compatibility, and Standards: The Economics of Horses, Penguins and Lemmings." In *Product Standardization and Competitive Strategy*, H. Landis Gabel, ed. Amsterdam: Elsevier Science Publishers B.V., 1987.

Farrell, Joseph, and Garth Saloner. "Standardization and Variety." In *Economics Letters* 20 (1986): 71-74.

Farrell, Joseph, and Garth Saloner. "Standardization, Compatibility, and Innovation." In *RAND Journal of Economics* 16, no. 1 (1985): 70-83.

Federal Trade Commission. *Standards and Certification: Final Staff Report*. Washington, D.C.: U.S. Government Printing Office, 1983.

Finley, Foster, and Don Swann. *ISO 9000 Survey Results*. Deloitte & Touche Management Consulting, January 1994.

Finley, Foster. *ISO 9000: Investment or Expense?* Deloitte & Touche Management Consulting, 1993.

Funabashi, Yoichi; Michel Oksenberg, and Heinrich Weiss. "An Emerging China in a World of Interdependence. A Report to the Trilateral Commission." In *The Triangle Papers: 45*. New York: The Trilateral Commission, 1994.

Gabel, H. Landis, ed. *Product Standardization and Competitive Strategy*. Amsterdam: Elsevier Science Publishers B.V., 1987.

Greenstein, Shane M. "Invisible Hands and Visible Advisors: An Economic Interpretation of Standardization." In *Journal of the American Society for Information Science* 43, no. 8 (1992): 538-549.

Grieco, Joseph. *Cooperation Among Nations: Europe, America, and Non-Tariff Barriers to Trade*. Ithaca, N.Y.: Cornell University Press, 1990.

Hamilton, Robert W. "The Role of Nongovernmental Standards in the Development of Mandatory Federal Standards Affecting Safety or Health." In *Texas Law Review* 56, no. 8 (1978): 1329-1484.

Health Industry Manufacturers Association. *EU-U.S. Mutual Recognition Agreements (MRAs): Key Issues for the Medical Device Industry*. March 29, 1994.

Hemenway, David. *Industry-Wide Voluntary Product Standards*. Cambridge, Mass.: Ballinger Publishing Company, 1975.

Hergert, Michael. "Technical Standards and Competition in the Microcomputer Industry." In *Product Standardization and Competitive Strategy*, H. Landis Gabel, ed. Amsterdam: Elsevier Science Publishers B.V., 1987.

Hertz, Harry S., and Curt W. Reimann. "The Malcolm Baldrige National Quality Award and ISO 9000 Registration." In *ASTM Standardization News* (November 1993).

Hiraiwa Commission. *Regarding Economic Structural Reform: Proposal.* Advisory Group for Economic Structural Reform. December 1993.

Hudec, Robert E. *The GATT Legal System and World Trade Diplomacy.* New York: Praeger, 1975. In Patrick Low. *Trading Free: The GATT and U.S. Trade Policy.* New York: The Twentieth Century Fund, 1993.

Hudec, Robert E. *Developing Countries in the GATT System.* Thames Essay No. 50. Aldershot, U.K.: Gower for the Trade Policy Research Centre, 1987. In Patrick Low. *Trading Free: The GATT and U.S. Trade Policy.* New York: The Twentieth Century Fund, 1993.

Hufbauer, Gary Clyde. *Wither APEC?* Presented at the Second Conference on "APEC: NAFTA/ ASEAN/SAARC." August 29 - September 1, 1994, in Nusa Dua, Bali, Indonesia. Washington, D.C.: IIE, 1994.

Hufbauer, Gary Clyde, and Jeffrey J. Schott. *NAFTA: An Assessment.* Revised edition. Institute for International Economics. Washington, D.C.: IIE, October 1993.

Hufbauer, Gary Clyde, and Joanna Shelton Erb. *Subsidies in International Trade.* Washington, D.C.: Institute for International Economics, 1984. In Patrick Low. *Trading Free: The GATT and U.S. Trade Policy.* New York: The Twentieth Century Fund, 1993.

Hufbauer, Gary C. *The Free Trade Debate. Twentieth Century Fund Task Force on the Future of American Trade Policy.* Table 2.10, p. 74. New York: Twentieth Century Fund, 1989. In Patrick Low. *Trading Free: The GATT and U.S. Trade Policy.* New York: The Twentieth Century Fund, 1993.

Hufbauer, Gary C., and Jeffrey J. Schott. *North American Free Trade: Issues and Recommendations.* Institute for International Economics. Washington, D.C.: IIE, 1992.

Hufbauer, Gary C., and Jeffrey J. Schott. "Options for a Hemispheric Trade Order." In *The University of Miami Inter-American Law Review* 22, no. 2-3 (1991): 261-296.

Hufbauer, Gary Clyde; Jeffrey J. Schott, and Kimberly Ann Elliott. *Economic Sanctions Reconsidered.* Institute for International Economics. Washington, D.C.: IIE, 1990.

Hughes, Kirsty S., ed. *European Competitiveness.* Great Britain: Cambridge University Press, 1993.

Hyer, Charles. "Performance Review Institute." In *TMO Update.* Ridgefield, Conn.: The Marley Organization, 1994.

Hyer, Charles, ed. *Directory of Professional/Trade Organization Laboratory Accreditation/Designation Programs.* Gaithersburg, Md.: NIST, 1992.

Hyer, Charles W., ed. *Directory of State and Local Government Laboratory Accreditation/Designation Programs.* NIST Special Publication 815. National Institute of Standards and Technology, U.S. Department of Commerce. Washington, D.C.: U.S. Government Printing Office, 1991.

Institute of Electrical and Electronics Engineers. *Standards Catalog.* Piscataway, N.J.: IEEE, 1994.

Institute of Electrical and Electronics Engineers. *Annual Activities Report.* Piscataway, N.J.: IEEE, 1993.

International Bank for Reconstruction and Development. *Development in Practice, East Asia's Trade and Investment, Regional and Global Gains from Liberalization.* Washington, D.C.: IBRD, 1994.

International Organization for Standardization. Council Committee on Conformity Assessment. *Information on CASCO.* Brochure, 1993.

International Organization for Standardization. *Certification and Related Activities: Assessment and Verification of Conformity to Standards and Technical Specifications.* Geneva: ISO, 1992.

International Organization for Standardization. *Compendium of Conformity Assessment Documents.* Geneva: ISO, 1991.

ISO 9000 Central Secretariat. *ISO 9000 International Standards for Quality Management - Compendium.* 2nd edition. Geneva: International Organization for Standardization, 1992.

Jackson, John H. *World Trade and the Law of GATT.* Indianapolis, Ind.: Bobbs-Merrill, 1969: p. 37.

In Patrick Low. *Trading Free: The GATT and U.S. Trade Policy.* New York: The Twentieth Century Fund, 1993.

Jackson, John H. "The Birth of the GATT-MTN System: A Constitutional Appraisal." In *Law and Policy in International Business* 12, no. 21 (1983): 21. In Patrick Low. *Trading Free: The GATT and U.S. Trade Policy.* New York: The Twentieth Century Fund, 1993.

Jackson, John H. *The World Trading System: Law and Policy of International Relations.* Cambridge, Mass.: MIT Press, 1989. In Patrick Low. *Trading Free: The GATT and U.S. Trade Policy.* New York: The Twentieth Century Fund, 1993.

Jackson, John H., and William J. Davey. *Legal Problems of International Economic Relations.* Second edition. St. Paul, Minn.: West Publishing Co., 1986.

Japan, Government of. *Government of Japan's Deregulation Package.* As explained in June 1994 Incoming Telegram to Department of Commerce from American Embassy-Tokyo (Action Copy and Facsimile).

Japan Accreditation Board for Quality System Registration. *Outline of the Japan Accreditation Board for Quality System Registration.* Tokyo, Japan: JAB, 1994.

Japanese Standards Association. *Industrial Standardization in Japan, 1994.* Japanese Industrial Standards Committee. Japan: Japanese Standards Association, 1994.

Jensen, Richard, and Marie Thursby. *Patent Races, Product Standards, and International Competition.* Unpublished draft. July 1994.

Jussawalla, Meheroo, ed. "United States-Japan Trade in Telecommunications: Conflict and Compromise." In *Contributions in Economics and Economic History*, Series Number 145. Dominick Salvatore, ed. Westport, Conn.: Greenwood Press, 1993.

Kantor, Michael. *The Importance of the Uruguay Round.* Statement to the House Ways and Means Committee. July 14, 1994.

Kasman, Bruce. "Recent U.S. Export Performance in the Developing World." In *Federal Reserve Bank of New York Quarterly Review* 17, no. 4 (1992-3): 64-74.

Kindleberger, Charles P. "Standards as Public, Collective and Private Goods." In *Kyklos* 36 (1983): 377-396.

Klien, Pamela. "NSF Wins Court Fight." In *The Ann Arbor (Mi.) News.* October 24, 1980, sec. C.

Kolb, John, and Steven S. Ross. *Product Safety and Liability: A Desk Reference.* New York: McGraw-Hill, 1980.

Laird, Sam, and Alexander Yeats. *Quantitative Methods for Trade Barrier Analysis.* London: MacMillan, 1990.

Lawler, Mary Anne. Memorandum. U.S. Technical Advisory Group to JTC-1, June 1994.

Lecraw, Donald J. "Japanese Standards: A Barrier to Trade?" In *Product Standardization and Competitive Strategy*, H. Landis Gabel, ed. Amsterdam: Elsevier Science Publishers B.V., 1987.

Lecraw, Donald J. "Some Economic Effects of Standards." In *Applied Economics* 16 (1984): 507-522.

Lehr, William. "Standardization: Understanding the Process." *Journal of the American Society for Information Science* 43, no. 8 (1992): 550-555.

Lincoln, Edward J. *Japan's Unequal Trade.* Washington, D.C.: Brookings Institution, 1990.

Lincoln, Edward J. "Japan's Role in Asia-Pacific Cooperation: Dimensions, Prospects, and Problems." In *Journal of NorthEast Asian Studies* (Winter 1989): 3-23.

Lincoln, Edward J. *Japan's Economic Role in Northeast Asia.* The Asia Society and University Press of America. Lanham, Md.: University Press of America, Inc., 1987.

Link, Albert N. "Evaluating the Advanced Technology Program: A Preliminary Assessment of Economic Impacts." In *International Journal of Technology Management* 8, no. 6-7-8 (1993): 726-739.

Link, Albert N., and Gregory Tassey. "The Impact of Standards on Technology-Based Industries: The Case of Numerically Controlled Machine Tools in Automated Batch Manufacturing." In

Product Standardization and Competitive Strategy, H. Landis Gabel, ed. Amsterdam: Elsevier Science Publishers B.V., 1987.

Link, Albert N. "Market Structure and Voluntary Product Standards." In *Applied Economics* 15 (1983): 393-401.

Locke, John W. *Conformity Assessment—At What Level?* American Association for Laboratory Accreditation. Presented at the Joint ISO, ANSI, and ASQC ISO 9000 Forum Application Symposium, Washington, D.C., October 7, 1993.

Locke, John W. *Trends in U.S. and International Approaches to Accreditation of Laboratories and Testing and Certification*. Paper. American Association for Laboratory Accreditation, November 23, 1992.

Low, Patrick. *Trading Free: The GATT and U.S. Trade Policy*. A Twentieth Century Fund Book. New York: Twentieth Century Fund Press, 1993.

Low, Patrick, ed. *International Trade and the Environment*. Discussion Paper No. 159. Washington, D.C.: World Bank, 1992. In Patrick Low. *Trading Free: The GATT and U.S. Trade Policy*. New York: The Twentieth Century Fund, 1993.

Ludolph, Charles M. "Mutual Recognition Agreements, Part III: Summary of Negotiations at Midpoint" (unpublished draft, July 1994).

Ludolph, Charles M. "Mutual Recognition Agreements Part II." In *The European Report on Industry—Quality and Standards*. May/June 1994.

Ludolph, Charles M. "Mutual Recognition Agreements—Access to the European Union." In *The European Report on Industry: Quality and Standards*. Part of *The Complete European Digest* 3, no. II (March 1994). Atlanta, Ga.: SIMCOM, 1994.

Ludwiszewski, Raymond B. "'Green' the Language in the NAFTA: Reconciling Free Trade and Environmental Protection." In *The International Lawyer* 27, no. 3 (1993): 691-706.

Maddison, Angus. *The World Economy in the 20th Century*. Paris: OECD, 1989. In Patrick Low. *Trading Free: The GATT and U.S. Trade Policy*. New York: The Twentieth Century Fund, 1993.

Mahoney, Edward F. "The ISO 9000 Quality System Trade Barrier of Foundation for Continuous Quality Improvement." In *Ceramic Engineering and Science Proceedings* 14, no. 3-4 (1993): 36-44.

Manufacturers' Alliance for Productivity and Innovation. *The European Community's New Approach to Regulation of Product Standards and Quality Assurance (ISO 9000): What it means for U.S. Manufacturers*. ER-218. Washington, D.C.: MAPI, 1992.

Marash, Stanley A., and Donald W. Marquardt. "Quality, Standards, and Free Trade." In *Quality Progress* (May 1994): 27-30.

Martin, James, ed. *Telecommunications and the Computer: Third Edition*. Englewood Cliffs, N.J.: Prentice-Hall, 1990.

McClelland, Nina I., David A. Gregorka, and Betsy D. Carlton, "The Drinking Water Additives Program," *Environmental Science & Technology* 23, no. 1, 1989.

McGovern, Edmond. *International Trade Regulation*. Exeter: Globefield Press, 1986. In Patrick Low. *Trading Free: The GATT and U.S. Trade Policy*. New York: The Twentieth Century Fund, 1993.

Melo, Jaime de, and David Tarr. *A General Equilibrium Analysis of U.S. Foreign Trade Policy*. Cambridge, Mass.: MIT Press, 1992. In Patrick Low. *Trading Free: The GATT and U.S. Trade Policy*. New York: The Twentieth Century Fund, 1993.

Miller, Michael. Presentation to the Conference on New Developments in International Standards and Global Trade. Washington, D.C., March 30, 1994.

Miller, Michael J. *Hearing on International Standardization: The Federal Role*. Testimony before the U.S. House of Representatives Committee on Science, Space and Technology and Subcommittee on Science, Research and Technology. July 25, 1989.

Molka, Judith A. "Surrounded by Standards, There is a Simpler View." In *Journal of the American Society for Information Science* 43, no. 8 (1992): 526-530.

Mowry, Keith. *Conformity Assessment: An Extra Benefit from Standards.* Paper for Underwriters Laboratories, Inc. Washington, D.C.: Underwriters Laboratories, Inc., 1994.

Nadkarni, R. A. "ISO 9000: Quality Management Standards for Chemical and Process Industries." In *Analytical Chemistry* 65, no. 8 (1993): 387-395.

Nakhai, Behnam; and Joao S. Neves. "The Deming, Baldrige, and European Quality Awards." In *Quality Progress* (April 1994): 33-37.

National Academy of Engineering. Steering Committee on Product Liability and Innovation. *Product Liability and Innovation: Managing Risk in an Uncertain Environment.* Washington, D.C.: National Academy Press, 1994.

National Academy of Engineering. Committee on Technology Policy Options in a Global Economy. *Prospering in a Global Economy: Mastering a New Role.* Washington, D.C.: National Academy Press, 1993.

National Aerospace and Defense Contractors Accreditation Program. *Quality Accredited - Once and For All.* Information booklet. n.d.

National Fire Protection Association. *National Fire Protection Association: Fact Sheet.* Arlington, Va.: NFPA, n.d.

National Fire Protection Association. *The NFPA Standards-Making System.* Booklet. n.d.

National Fire Protection Association. *National Electrical Code 1993.* Quincy, Mass.: NFPA, 1992.

National Institute of Standards and Technology. *National Voluntary Laboratory Accreditation Program: Procedures and General Requirements.* NIST Handbook 150. Gaithersburg, Md.: NIST, 1994.

National Institute of Standards and Technology. *National Voluntary Laboratory Accreditation Program: Fee Schedule and Worksheets.* Brochure. 1994.

National Institute of Standards and Technology. "Establishment of the National Voluntary Conformity Assessment System Evaluation Program." In *Federal Register* 59, no. 78, (April 22, 1994).

National Institute of Standards and Technology. *NIST Budget Summary.* Available through Internet at gopher://gopher-server.nist.gov. August 30, 1994.

National Institute of Standards and Technology. "Weights and Measures." In *Guide to NIST.* Available through Internet at gopher://gopher-server.nist.gov. November 1993.

National Institute of Standards and Technology. *The Office of Standards Services.* Brochure. n.d.

National Research Council. Computer Science and Telecommunications Board. *Realizing the Information Future: The Internet and Beyond.* Washington, D.C.: National Academy Press, 1994.

National Research Council. Board on Telecommunications and Computer Applications. *Crossroads of Information Technology Standards.* Washington, D.C.: National Academy Press, 1990.

NSF International. *NSF International Corporate Brochure.* Ann Arbor, Mich.: NSF International, n.d.

Office of Management and Budget. *Federal Participation in the Development and Use of Voluntary Standards.* OMB Circular No. A-119 Revised. Leon E. Panetta, Director. Washington, D.C.: Executive Office of the President, October 20, 1993.

Office of Management and Budget. *Performance of Commercial Activities.* OMB Circular No. A-76 Revised. Washington, D.C.: Executive Office of the President, August 4, 1983.

Office of Management and Budget. *Circular No. A-119.* Washington, D.C.: Executive Office of the President, 1982.

Office of Trade and Investment Ombudsman, Japan. "OTO Responds to Complaints Regarding Import Procedures and Other Issues which Concern Opening of the Japanese Market." In *Guide to OTO.* Japan: OTO, n.d.

Organization for Economic Cooperation and Development. *Product Safety: Developing and Implementing Measures.* Paris: OECD, 1987.

Overman, JoAnne R. *GATT Standards Code Activities of the National Institute of Standards and Technology 1992.* Prepared for the National Institute of Standards and Technology, Technology Administration, U.S. Department of Commerce. Gaithersburg, Md.: NIST, 1993.

Pastor, Robert A. *Integration with Mexico: Options for U.S. Policy.* A Twentieth Century Fund Paper. New York: Twentieth Century Fund Press, 1993.

Patel, Surendra J., ed. *Technological Transformation in the Third World—Volume 1: Asia.* United Nations University. Great Britain: Avebury, 1993.

Peach, Robert W., ed. *The ISO 9000 Handbook.* 2nd edition. CEEM Information Services. Fairfax, Virginia: CEEM Information Services, 1994.

Pelkmans, Jacques, and Rita Beuter. "Standardization and Competitiveness: Private and Public Strategies in the EC Color TV Industry." European Institute of Public Administration. In *Product Standardization and Competitve Strategy*, H. Landis Gabel, ed. Amsterdam: Elsevier Science Publishers B.V., 1987.

Performance Review Institute. *National Aerospace and Defense Contractors Accreditation Program.* PRI, 1994.

Peyton, Donald L., and Charles W. Hyer. *Standards and Trade in the 1990s: A Source Book for Department of Defense Acquisition and Standardization Management and their Industrial Counterparts.* Defense Systems Management College. Washington, D.C.: U.S. Government Printing Office, 1993.

Quality. "International Standards: It's a Small World After All," In *Quality* (August 1986): special section.

Quality Systems Update. *Quality Systems Update* 4, no. 5 (May 1994).

Quality Systems Update. "DoD, NASA Officials: Thumbs Up on ISO 9000." In *Quality Systems Update* 4, no. 2 (February 1994).

Reddy, N. Mohan. "Technology, Standards, and Markets: A Market Institutionalization Perspective." In *Product Standardization and Competitive Strategy*, H. Landis Gabel, ed. Amsterdam: Elsevier Science Publishers B.V., 1987.

Reimann, Curt W., and Harry S. Hertz. "The Malcolm Baldrige National Quality Award and ISO 9000 Registration: Understanding Their Many Important Differences." In *ASTM Standardization News* (November 1993): 42-53.

Retlif Testing Laboratories. *U.S./EU Trade Negotiations: EMC Issues.* Memorandum, June 1994.

Rezendes, Victor S., and Jim Wells. *Metric Conversion: Future Progress Depends Upon Private Sector and Public Support.* A Report to Congressional Requesters GAO/RCED-94-23. U.S. General Accounting Office. Washington, D.C.: U.S. GAO, 1994.

Rountree, James E., ed. *Directory of DoC Staff Memberships on Outside Standards Committees.* National Institute of Standards and Technology, U.S. Department of Commerce. Gaithersburg, Md.: NIST, 1993.

Russell, Tony. "ISO 9000 and ISO/IEC Guide 25: Quality Systems Certification and Laboratory Accreditation—Competitive or Complimentary Activities?" Paper presented at the ISO 9000 Forum, Gold Coast, November, 1992. In *NATA News* (December 1992): 16-19.

Saito, Tadashi. "Product Liability Reform in Japan." In *Japan Economic Institute Report* no. 3A (January 21). Washington, D.C.: JEI, 1994.

Saito, Tadashi. "Role of Japan's Office of Trade and Investment Ombudsman." In *Japan Economic Institute Report* no. 35A (September 24). Washington, D.C.: JEI, 1993.

Saunders, Greg, et al. *Road Map for Milspec Reform: Integrating Commercial and Military Manufacturing.* Report of the Working Group on Military Specifications and Standards. Washington, D.C.: The Center for Strategic and International Studies, 1993.

Schott, Jeffrey J. *The Uruguay Round: An Assessment.* Washington, D.C.: Institute for International Economics, 1994.

Schott, Jeffrey J., ed. *Free Trade Areas and U.S. Trade Policy.* Washington, D.C.: Institute for

International Economics, 1989. In Patrick Low. *Trading Free: The GATT and U.S. Trade Policy*. New York: The Twentieth Century Fund, 1993.

Services of the European Commission. *Report on United States Barriers to Trade and Investment, 1994*. Brussels: Services of the European Commission, April 1994.

Simson, Bert G. *Conformity Assessment Workshop on Electromagnetic Compatibility*. National Institute of Standards and Technology. Gaithersburg, Md.: NIST, June 1991.

Siwek, Stephen E., and Harold W. Furchtgott-Roth. *International Trade in Computer Software*. Westport, Conn.: Quorum Books, 1993.

Society of Automotive Engineers. *1994 Report to Ground Vehicle Industry*. 1994 SAE Cooperative Engineering Program.

Spickernell, D. G. "The Role of International Standards in Removing Barriers to Trade and Assisting Developing Countries." In *International Journal of Technology Management* 1, no. 1/2 (1986): 197-208.

Stanger, David. Secretary General, European Organization for Testing and Certification. *EOTC*. Presentation at ANSI Annual Conference, Washington, D.C., March 4, 1994.

Steinberg, Richard H. "The Uruguay Round: A Legal Analysis of the Final Act." In *International Quarterly* 6, no. 2 (1994): 1-97.

Sterling, Cristopher H. "The FCC and Changing Technological Standards." In *Journal of Communication* 32 (1982): 137-147.

Sullivan, Charles D. *Standards and Standardization: Basic Principles and Applications*. New York: Marcel Dekker, Inc., 1983.

Tassey, Gregory. *Technology Infrastructure and Competitive Position*. Norwell, Mass.: Kluwer Academic Publishers, 1992.

Thayer, Ann M. "Chemical Companies See Beneficial Results From ISO 9000 Registration". In *Chemical & Engineering News* (April 25, 1994).

Thurwachter, Todd. "Japan's Non-Tariff Barriers: Case Studies from Files of U.S. and Foreign Commercial Service." In *The Journal of the ACCJ* (November 1988): 33-49.

Tokyo Round. General Agreement on Tariffs and Trade. *Agreement on Technical Barriers to Trade*. Tokyo Round Agreements, 1979.

Toth, Robert B., ed. *Standards Activities of Organizations in the United States*. NIST Special Publication 806. National Institute of Standards and Technology. U.S. Department of Commerce. Washington, D.C.: U.S. Government Printing Office, 1991.

Toth, Robert B. *Standards Management: A Handbook for Profits*. New York: American National Standards Institute, Inc., 1990.

Tronel, Lucien. *Conformity Assessment Activities and International Standardization in E.C. Countries*. Paris: Association Francaise de Normalisation, October 1993.

Tussie, Diana. *The Less Developed Countries and the World Trading System: A Challenge to the GATT*. New York: St. Martin's Press, 1987. In Patrick Low. *Trading Free: The GATT and U.S. Trade Policy*. New York: The Twentieth Century Fund, 1993.

Under Secretary of Defense for Acquisition and Technology, Office of the. U.S. Department of Defense. *Report of the Process Action Team on Military Specifications and Standards*. Washington, D.C.: U.S. DoD, April 1994.

Underwriters Laboratories. *An Overview of Underwriters Laboratories*. Brochure. Northbrook, Ill.: UL, 1993.

Underwriters Laboratories. *Underwriters Laboratories, Inc. Annual Report 1993*. Northbrook, Ill.: UL, 1993.

Underwriters Laboratories. *More Than a Mark*. Information booklet. Northbrook, Ill.: UL, 1993.

United Nations Council on Trade and Development (UNCTD). *Problems of Protectionism and Structural Adjustment: Restrictions on Trade*. TD-B-1126. Geneva: UNCTD, 1987.

Uruguay Round. General Agreement on Tariffs and Trade. *Agreement on the Application of Sanitary and Phytosanitary Measures*. MTN/FA II-A1A-4. 1993.

U.S. Bureau of the Census. Economics and Statistics Administration. *Service Annual Survey: 1992.* BS/92, Current Business Reports. Washington, D.C.: U.S. Department of Commerce, 1992.

U.S. Bureau of the Census. Economics and Statistics Administration. *1987 Census of Service Industries.* SC87-S-4 Subject Series, Miscellaneous Subjects. Washington, D.C.: U.S. Department of Commerce, 1991.

U.S. Bureau of the Census. Economics and Statistics Administration. *Service Annual Survey: 1991.* BS/91-1, Current Business Reports. Washington, D.C.: U.S. Department of Commerce, 1991.

U.S. Congress. *North American Free Trade Agreement Implementation Act.* Public Law 103-182. 103rd Congress, 2nd Session. December 8, 1993. Washington, D.C.: U.S. Government Printing Office, 1993.

U.S. Congress. *Fastener Quality Act.* Public Law 101-592. 101st Congress, 2nd session. November 16, 1990.

U.S. Congress. House. *Department of Commerce Appropriations Act, 1995.* 103rd Congress, 2nd session, H.R. 4603.

U.S. Congress. House. Committee on Ways and Means. *Overview and Compilation of U.S. Trade Statutes.* 1993 edition. 103rd Congress. Washington, D.C.: U.S. Government Printing Office, 1993.

U.S. Congress. House. Committee on Science, Space, and Technology. *International Standards and Trade, Hearing No. 112.* 102nd Congress, 2nd session, 1992.

U.S. Congress. House. Committee on Armed Services. *DoD Military Specifications and Standards, HASC No. 102-75.* 102nd Congress, 2nd session, 1992.

U.S. Congress. House. Committee on Science, Space, and Technology. *International Standards, Hearing No. 135.* 101st Congress, 2nd session, 1990.

U.S. Congress. House. Committee on Small Business. *European Community Approach to Testing and Certification: Should the U.S. Government Play a Role? Hearing No. 56.* 101st Congress, 2nd session, 1990.

U.S. Congress. House. Committee on Science, Space, and Technology. *The Federal Role in International Testing, Certification and Quality Assurance, Hearing No. 145.* 101st Congress, 2nd session, 1990.

U.S. Congress. House. Committee on Science, Space, and Technology. *International Standardization: The Federal Role, Hearing No. 52.* 101st Congress, 1st session, 1989.

U.S. Congress. House. *Trade Agreements Act of 1979.* 96th Congress, 1st session, 1979.

U.S. Congress. Office of Technology Assessment. *Development Assistance, Export Promotion, and Environmental Technology—Background Paper.* OTA-BP-ITE-107. Washington, D.C.: U.S. Government Printing Office, 1993.

U.S. Congress. Office of Technology Assessment. *U.S. Telecommunications Services in European Markets.* OTA-TCT-548. Washington, D.C.: U.S. Government Printing Office, 1993.

U.S. Congress. Office of Technology Assessment. *Global Standards: Building Blocks for the Future.* TCT-512. Washington, D.C.: U.S. Government Printing Office, 1992.

U.S. Congress. Senate. Committee on the Judiciary. *Voluntary Standards and Accreditation Act of 1977.* Hearings on S. 825, 95th Congress, 1st session, 1977.

U.S. Department of Commerce. International Trade Administration. *EC Product Standards Under the Internal Market Program.* Washington, D.C.: U.S. DoC, 1993.

U.S. Department of Commerce. International Trade Administration. *EC Testing and Certification Procedures Under the Internal Market Program.* Washington, D.C.: U.S. DoC, 1993.

U.S. Department of Commerce. International Trade Administration. *Japanese Regulations, Standards, Quality Marks, and Certification Systems.* Destination Japan (2nd. ed.). 82-95. Washington, D.C.: U.S. DoC, 1994.

U.S. Department of Commerce. Memorandum. *Third Triennial Report to the Office of Management and Budget on the Implementation of OMB Circular A-119*, February 26, 1992. Washington, D.C.: U.S. DoC, 1992.

U.S. Department of Commerce. Office of Japan Affiars. *Fact Sheet: Standards, Certification and Testing.* ETRD 2705.

U.S. Department of Commerce. Technology Administration. NIST. *National Voluntary Laboratory Accreditation Program: Fee Schedule and Worksheets, Testing Laboratories, January 1994.* Gaithersburg, Md.: National Institute of Standards and Technology, 1994.

U.S. Department of Commerce. Technology Administration. NIST. *Directory of DOC Staff Memberships on Outside Standards Committees. January 1994.* Gaithersburg, Md.: National Institute of Standards and Technology, 1994.

U.S. Department of Commerce. Technology Administration. NIST. *Advanced Technology Program: Proposal Preparation Kit, February 1994.* Gaithersburg, Md.: National Institute of Standards and Technology, 1994.

U.S. Department of Commerce. Technology Administration. *NVLAP, National Voluntary Laboratory Accreditation Program, 1994 Directory.* Special Publication 810. Gaithersburg, Md.: National Institute of Standards and Technology, 1994.

U.S. Department of Commerce. Technology Administration. NIST. *National Voluntary Laboratory Accreditation Program: Procedures and General Requirements.* NIST Handbook 150. March 1994. Washington, D.C.: U.S. Government Printing Office, 1994.

U.S. Department of Defense. *Qualified Manufacturers List (QML): Capturing Commercial Technology for Microelectronics.* Pamphlet. Washington, D.C.: U.S. DoD, n.d.

U.S. Environmental Protection Agency. Jeanne Hankins, ed. *Final Report of the Committee on National Accreditation of Environmental Laboratories.* September 1992. Washington, D.C.: EPA, 1992.

U.S. General Accounting Office. *Management Practices: U.S. Companies Improve Performance through Quality Efforts. A Report to the Honorable Donald Ritter, U.S. House of Representatives.* National Security and International Affairs Division and General Government Division. Washington, D.C.: U.S. GAO, 1991.

U.S. General Accounting Office. *Laboratory Accreditation: Requirements Vary Throughout the Federal Government. A Report to the Chairman, Committee on Science, Space, and Technology. U.S. House of Representatives.* Resources, Community, and Economic Development. GAO/RCED-89-102. Washington, D.C.: U.S. GAO, 1989.

U.S. International Trade Commission. *Potential Impact on the U.S. Economy and Selected Industries of the North American Free Trade Agreement.* Publication 2596. January 1993.

U.S. International Trade Commission. *Development Assistance in East Asia.* Washington, D.C.: U.S. Government Printing Office, 1993.

U.S. International Trade Commission. *Global Competitiveness of U.S. Advanced-Technology Industries: Cellular Communications.* Publication 2646. Investigation No. 332-329. 1993.

U.S. International Trade Commission. *The Effects of Greater Economic Integration within the European Community on the United States: Fifth Followup Report.* Publication 2628. Investigation No. 332-267. 1993.

U.S. International Trade Commission. *The Effects of Greater Economic Integration within the European Community on the United States: Fourth Followup Report.* Publication 2501. Investigation No. 332-267. 1992.

U.S. International Trade Commission. Office of Economics. *East Asia: Regional Economic Integration and Implications for the United States.* Publication 2621. May 1993.

U.S. International Trade Commission. Office of Industries. "Technical Standards and International Competition: The Case of Cellular Communications." In *Industry Trade and Technology Review* (October 1993): 11-16.

U.S. Trade Representative, Office of the. *Uruguay Round: Final Texts of the GATT Uruguay Round Agreements.* MTN/FA II-AIA-6 Washington, D.C.: U.S. Government Printing Office, 1994.

U.S. Trade Representative, Office of the. *1994 National Trade Estimate Report on Foreign Trade Barriers.* Various country reports. Washington, D.C.: USTR, 1994.

U.S. Trade Representative, Office of the. *1993 Annual Report*. Washington, D.C.: USTR, 1994.

U.S. Trade Representative, Office of the, Office of the Chief Economist. *U.S. Exports Create High-Wage Employment*. Washington, D.C.: USTR, June 24, 1992.

U.S. Trade Representative, Office of the, Office of Public Affairs. Facsimile describing the USTR and its functions.

Visiting Committee on Advanced Technology of the National Institute of Standards and Technology. *International Standards Issues: A Statement to the Secretary of Commerce*. July 28, 1993.

Vogt, Donna U. *Sanitary and Phytosanitary Safety Standards for Foods in the GATT Uruguay Round Accords*. Congressional Research Service (94-512-SPR), Library of Congress, June 21, 1994. Washington, D.C.: CRS, 1994.

Walters, David. *Comparison of Export Sector Wages to Overall Sector Wages*. Washington, D.C.: Office of the U.S. Trade Representative, June 1992.

WECC International Symposium. *Calibration and the European Market: Current and Future Policy for the 1990s*. Proceedings. May 1993. Royal Crown Hotel, Brussels.

Weiss, Martin, and Carl Cargill. "Consortia in the Standards Development Process." In *Journal of the American Society for Information Science* 43, no. 8 (1992): 559-565.

Weiss, Martin B. H., and Marvin Sirbu. "Technological Choice in Voluntary Standards Committees: An Empirical Analysis." In *Economics of Innovation and New Technology* 1 (1990): 111-133.

White House. *Economic Report of the President, January 1989*. Table B-58, p. 373. Washington, D.C.: U.S. Government Printing Office, 1989. In Patrick Low. *Trading Free: The GATT and U.S. Trade Policy*. New York: The Twentieth Century Fund, 1993.

White House. *Economic Report of the President, January 1993*. Table B-1, p. 348. Washington, D.C.: U.S. Government Printing Office, 1993. In Patrick Low. *Trading Free: The GATT and U.S. Trade Policy*. New York: The Twentieth Century Fund, 1993.

White House. "Trade Agreements Resulting from the Uruguay Round of Multilateral Trade Negotiations." Memorandum. In *Federal Register* 58, no. 242 (December 1993): 67269-67270.

Wilson, John Sullivan. "The U.S. Performance in Advanced Technology Trade: 1982-93 (est.)." In *Challenge* (January 1994).

Wilson, John Sullivan. *Standards, Conformity Assessment, and Trade: New Developments and the Asia Pacific Economic Cooperation (APEC) Forum*. Paper presented to The Second Conference on "APEC: NAFTA/ASEAN/SAARC." August 29–September 1, 1994, Bali, Indonesia: n.p., 1994.

Wilson, John S. *The U.S. Government Trade Policy Response to Japanese Competition in Semiconductors: 1982-87*. September 1987.

Woodcock, Hank. "Nationally Recognized Testing Laboratories." In *Job Safety & Health Quarterly*. Summer 1993.

World Bank. *East Asia's Trade and Investment: Regional and Global Gains from Liberalization*. Washington, D.C.: World Bank, 1994.

World Bank. *World Development Report 1987*. New York: Oxford University Press, 1987.

Wright, Don R., and Mary Saunders. "Status of EC Regulatory Harmonization Under the 'Old Approach.'" In *Business America* (March 1993): 32-35.

Index